Gerd Jürgen Winterlik

Unconventional correlated systems

Gerd Jürgen Winterlik

Unconventional correlated systems

Theory vs. experiment; prerequisites for successful material design

Südwestdeutscher Verlag für Hochschulschriften

Impressum / Imprint

Bibliografische Information der Deutschen Nationalbibliothek: Die Deutsche Nationalbibliothek verzeichnet diese Publikation in der Deutschen Nationalbibliografie; detaillierte bibliografische Daten sind im Internet über http://dnb.d-nb.de abrufbar.

Alle in diesem Buch genannten Marken und Produktnamen unterliegen warenzeichen-, marken- oder patentrechtlichem Schutz bzw. sind Warenzeichen oder eingetragene Warenzeichen der jeweiligen Inhaber. Die Wiedergabe von Marken, Produktnamen, Gebrauchsnamen, Handelsnamen, Warenbezeichnungen u.s.w. in diesem Werk berechtigt auch ohne besondere Kennzeichnung nicht zu der Annahme, dass solche Namen im Sinne der Warenzeichen- und Markenschutzgesetzgebung als frei zu betrachten wären und daher von jedermann benutzt werden dürften.

Bibliographic information published by the Deutsche Nationalbibliothek: The Deutsche Nationalbibliothek lists this publication in the Deutsche Nationalbibliografie; detailed bibliographic data are available in the Internet at http://dnb.d-nb.de.

Any brand names and product names mentioned in this book are subject to trademark, brand or patent protection and are trademarks or registered trademarks of their respective holders. The use of brand names, product names, common names, trade names, product descriptions etc. even without a particular marking in this work is in no way to be construed to mean that such names may be regarded as unrestricted in respect of trademark and brand protection legislation and could thus be used by anyone.

Verlag / Publisher:
Südwestdeutscher Verlag für Hochschulschriften
ist ein Imprint der / is a trademark of
OmniScriptum GmbH & Co. KG
Heinrich-Böcking-Str. 6-8, 66121 Saarbrücken, Deutschland / Germany
Email: info@svh-verlag.de

Herstellung: siehe letzte Seite /
Printed at: see last page
ISBN: 978-3-8381-1591-7

Zugl. / Approved by: Mainz, Johannes Gutenberg-Universität, Dissertation, 2009

Copyright © 2010 OmniScriptum GmbH & Co. KG
Alle Rechte vorbehalten. / All rights reserved. Saarbrücken 2010

Contents

1 Introduction — 3

2 Publications — 7

3 Challenging the prediction of anionogenic ferromagnetism for Rb_4O_6 — 9

4 Challenge of magnetism in strongly correlated open-shell 2p systems — 13

5 Exotic magnetism in the alkali sesquoxides Rb_4O_6 and Cs_4O_6 — 21

6 Electronic and structural properties of palladium-based Heusler superconductors — 33

7 Superconductivity in palladium-based Heusler compounds — 41

8 Ni-based superconductor: Heusler compound $ZrNi_2Ga$ — 61

9 Rational design of a novel noncentrosymmetric superconductor — 79

10 Structural, electronic, and magnetic properties of tetragonal $Mn_{3-x}Ga$: Experiments and first-principles calculations — 89

11 Summary — 115

1 Introduction

The class of strongly correlated systems encompasses a wide area of materials that often exhibit extraordinary electronic and magnetic properties such as metal-insulator transitions, exotic forms of magnetism, or superconductivity. At present correlated systems are in the center of attraction within the research areas of materials science and condensed matter physics because their exceptional properties are highly important for fundamental research and in many cases useful for technological applications.

The most popular correlated systems are the high-temperature superconductors [1], which are up to the present in the center of attraction, particularly due to the very recent discovery of a new class of Fe-based high-T_c superconductors [2]. Various transition metal oxides are correlated systems, which exhibit a large variety of interesting properties. The manganites with perovskite structure such as $La_{1-x}Ca_xMnO_3$ or $La_{1-x}Sr_xMnO_3$, which are structurally related to the cuprate superconductors, exhibit a colossal magnetoresistance (CMR), which allows massive changes of the electrical resistivity in the presence of a magnetic field [3, 4]. CMR materials are useful for the development of novel technologies such as read-write heads for magnetic storage and spintronics.

Electronic correlations are the effects beyond the Hartree-Fock theory. Within this approximation, the antisymmetric wave function is described by a single Slater determinant, which does not consider Coulomb correlation. This type of correlation is one major aspect that has to be addressed in correlated systems, which typically have open d- or f-electron shells. The electronic correlations in these systems lead to a narrowing of their electronic bands. This fact complicates the theoretical description of such materials by simple one-electron theories such as the local density approximation (LDA) of density functional theory or Hartree-Fock theory. The seemingly simple material NiO has an open $3d$-shell and is thus expected to be a metal. Due to electronic correlations between d-electrons, however, NiO is insulating [5].

This work comprises different unconventional correlated systems. Each of these systems has exceptional properties. The seemingly simple alkali sesquioxides (Chapters 3-5 in this work) demonstrate that electronic correlations are of major importance not only for d- and f-electron systems but also for $2p$-compounds. Heusler compounds are usually magnetic and widely known for their potential with respect to application in spintronic devices. The term "unconventional" pertaining to the superconducting Heusler compounds (Chapters 6-9 in this work) does not mean "Unconventional Superconductivity" [6] but rather that certain Heusler compounds with particular features in their electronic structure can be predicted to become superconducting. The magnetic correlations between Mn atoms at different crystallographic sites in the tetragonally distorted Heusler compound Mn_3Ga (Chapter 10) lead to magnetic properties that are of special importance for spintronics applications.

Chapters 3-5 of this work are about the open shell compounds Rb_4O_6 and Cs_4O_6. These mixed valent compounds contain oxygen in two different modifications: the closed-shell peroxide anion is nonmagnetic, whereas the hyperoxide anion contains an unpaired electrons in an antibonding π^*-orbital. Due to this electron magnetic ordering is rendered possible. In contrast to theoretical predictions [7], which suggested half-metallic ferromagnetism for Rb_4O_6, dominating antiferromagnetic interactions were found in the experiment [8]. Besides a symmetry reduction due to the mixed valency, strong electronic correlations of this highly molecular system determine its properties [9]; it is a magnetically frustrated insulator. The corresponding Cs_4O_6 was found to show similar properties [10].

Correlation of electrons can lead to magnetic interactions or to superconductivity and even to coexistence of magnetic ordering with the superconducting state. Chapters 6-9 of this work are about intermetallic Heusler superconductors [11–13]. All of these superconductors were rationally designed using the van Hove scenario as a working recipe [14, 15]. A saddle point in the energy dispersion curve of a solid leads to a van Hove singularity in the density of states. In the Ni-based and Pd-based Heusler superconductors presented in this work this sort of a valence instability occurs at the high-symmetry L point and coincides or nearly coincides with the Fermi level. The compounds escape the high density of states at the Fermi energy through a transition into the correlated superconducting state.

Another form of correlations is found in the binary Heusler-type compound Mn_3Ga. The cubic DO_3 Heusler-type phase of this compound with its 24 valence electrons was predicted to be a 100% spin-polarized half-metallic completely compensated ferrimagnet [16]. This prediction was made on the basis of the so-called Kübler rule for Heusler compounds. According to this rule, Mn at the Y position in a Heusler compound X_2YZ tends to a high localization. At this position Mn may be described as Mn^{3+} with a magnetic moment of approximately 4 μ_B. Chapter 10 of this work is about the tetragonally distorted ferrimagnetic DO_{22} phase of Mn_3Ga. This hard-magnetic modification is technologically useful for spin torque transfer applications [17–19]. The phase exhibits two different crystallographic sites that are occupied by Mn atoms and

1. Introduction

can thus be written as Mn_2MnGa. The competition between the mainly itinerant moments of the Mn atoms at the Wyckoff position $4d$ and the localized moments of the Mn atoms at the Wyckoff position $2b$ leads to magnetic correlations. The antiferromagnetic orientation of these moments determines the compound to exhibit a resulting magnetic moment of approximately 1 μ_B per formula unit in a partially compensated ferrimagnetic configuration.

Each chapter of this work is self-contained and corresponds to a scientific publication of Chapter 2. The styles of the different chapters (arrangement, figures etc.) were intentionally kept as published in the respective journals in order to accentuate the autonomy of each chapter. My contributions to the publications were the syntheses of all Heusler superconductors and of the $Mn_{3-x}Ga$ series and almost all physical characterizations as well as interpretations of the experimental results for the alkali sesquioxides, the Heusler superconductors, and $Mn_{3-x}Ga$ aside from the specific heat measurement of the Heusler superconductor $ZrNi_2Ga$ and the low-temperature x-ray diffraction and extended x-ray absorption fine structure (EXAFS) measurements of Mn_3Ga. The electronic structure calculations were carried out by G. H. Fecher, J. Kübler, K. Doll, and L. M. Sandrastkii.

2 Publications

1. C. Peters, S. Pechtl, J. Stutz, K. Hebestreit, G. Honninger, K. G. Heumann, A. Schwarz, J. Winterlik and U. Platt
   Reactive and organic halogen species in three different European coastal environments
   Atmos. Chem. Phys. **5**, 3357 (2005).

2. B. Balke, J. Winterlik, G. H. Fecher and C. Felser
   Mn_3Ga, a compensated ferrimagnet with high Curie temperature and low magnetic moment for spin torque transfer applications
   Appl. Phys. Lett. **90**, 152504 (2007).

3. J. Winterlik, G. H. Fecher, C. Felser, C. Mühle and M. Jansen
   Challenging the Prediction of Anionogenic Ferromagnetism for Rb_4O_6
   J. Am. Chem. Soc. **129** (22), 6990 (2007).

4. J. Winterlik, G. H. Fecher and C. Felser
   Electronic and structural properties of palladium-based Heusler superconductors
   Solid State Commun. **145**, 475 (2008).

5. J. Winterlik, B. Balke, G. H. Fecher, C. Felser, M. C. M. Alves, F. Bernardi and J. Morais
   Structural, electronic, and magnetic properties of tetragonal $Mn_{3-x}Ga$: Experiments and first-principles calculations
   Phys. Rev. B **77**, 054406 (2008).

6. J. Winterlik, G. H. Fecher, C. Felser, M. Jourdan, K. Grube, F. Hardy, H. von Löhneysen, K. L. Holman, and R. J. Cava

Ni-based superconductor: Heusler compound $ZrNi_2Ga$
Phys. Rev. B **78**, 184506 (2008).

7. J. Winterlik, G. H. Fecher, C. A. Jenkins, C. Felser, C. Mühle, K. Doll, M. Jansen, L. M. Sandratskii, and J. Kübler
Challenge of Magnetism in Strongly Correlated Open-Shell 2p Systems
Phys. Rev. Lett. **102**, 016401 (2009).

8. J. Winterlik, G. H. Fecher, A. Thomas, and C. Felser
Superconductivity in Palladium-based Heusler compounds
Phys. Rev. B **79**, 064508 (2009).

9. T. Graf, G. H. Fecher, J. Barth, J. Winterlik, and C. Felser
Electronic structure and transport properties of the Heusler compound Co_2TiAl
J. Phys. D: Appl. Phys. **42**, 084003 (2009).

10. T. Graf, F. Casper, J. Winterlik, B. Balke, G. H. Fecher, and C. Felser
Crystal Structure of New Heusler Compounds
Z. Anorg. Allg. Chem. **635**, 0000 (2009).

11. J. Winterlik, G. H. Fecher, C. A. Jenkins, C. Felser, J. Kübler, C. Mühle, K. Doll, M. Jansen, T. Palasyuk, I. Trojan, S. Medvedev, M. I. Eremets, and F. Emmerling
Exotic magnetism in the alkali sesquioxides Rb_4O_6 and Cs_4O_6
Accepted for publication in Phys. Rev. B (2009).

12. J. Winterlik, G. H. Fecher, C. Felser
Rational Design of a Novel Noncentrosymmetric Superconductor
To be submitted to Angew. Chem. Int. Ed. (2009).

3 Challenging the prediction of anionogenic ferromagnetism for Rb_4O_6

The text of this chapter is identical with the following publication:
J. Winterlik, G. H. Fecher, C. Felser, C. Mühle and M. Jansen
J. Am. Chem. Soc. **129**, 6990 (2007).

3. Challenging the prediction of anionogenic ferromagnetism for Rb$_4$O$_6$

For a long time, rubidium sesquioxide Rb$_4$O$_6$ has been of particular interest because its black color indicates an electronic structure that strongly differs from related systems such as regular alkali hyperoxides or peroxides. According to previous investigations, this compound exhibits a mixed valency type of behavior that is caused by differently charged dioxygen molecules, indicating that the structure may be written as Rb$_4$(O$_2^-$)$_2$(O$_2^{2-}$) [20]. The structure type Pu$_2$C$_3$ and the space group $I\bar{4}3d$, proposed by Helms and Klemm [21], as well as the presence of both peroxide and hyperoxide anions, have been verified by elastic and inelastic neutron scattering studies [22]. Attema *et al.* [7] performed density functional calculations within the local spin density approximation (LSDA), which suggest that Rb$_4$O$_6$ should be a half-metallic ferromagnet with the magnetic moment carried by the hyperoxide anions.

In the present work, magnetic measurements were performed to investigate this prediction of half-metallic ferromagnetism in Rb$_4$O$_6$. For the present experimental study, Rb$_4$O$_6$ samples were synthesized by the solid state reaction of RbO$_2$ and Rb$_2$O in a stoichiometric ratio [20, 23]. Purity and identity of the samples was confirmed by X-ray powder diffraction.

The magnetic properties of Rb$_4$O$_6$ were measured using a superconducting quantum interference device (SQUID, Quantum Design MPMS-XL5). Samples of approximately 100 mg, fused in Suprasil tubes under a helium atmosphere, were used for the analysis. Figures 3.1(a) and (b) display the magnetization data in the temperature range from 1.8 to 15 K. The sample was first cooled to a temperature of 1.8 K without applying a magnetic field. After applying an induction field $\mu_0 H$, the magnetization was then measured as the sample was heated from 1.8 to 15 K [zero-field-cooled (ZFC) modus]. Directly afterward, the measurements were performed in the same field as the temperature was again lowered down to 1.8 K [field-cooled (FC) measurements]. The magnetic phase transition was also examined using different strengths of the induction field $\mu_0 H$ that ranged from 2 mT to 5 T. The high temperature behavior of Rb$_4$O$_6$ (ZFC) was recorded as the temperature was varied between 100 and 300 K [Figure 3.1(c)].

From Figure 3.1, it is clear that the magnetization versus temperature curves exhibit hysteresis. This implies that the measurements are not reversible if they are started from a zero-field-cooled state. Such irreversibility between the FC and the ZFC measurements is typical for magnetically frustrated systems, for example, this effect is well-known to occur in spin glasses. From the high-temperature data, an effective magnetic moment of $m = 1.83$ μ_B per hyperoxide unit can be deduced applying the Curie-Weiss law, on the basis of molecular field theory (MFT). This value is in fair agreement with 1.73 μ_B, which is expected from MFT using the spin-only approximation. An analysis of the high-temperature data yields a negative paramagnetic transition temperature $\Theta_p = -6.9$ K, which is a characteristic signature of an antiferromagnetic type of interaction.

3. Challenging the prediction of anionogenic ferromagnetism for Rb$_4$O$_6$

No direct indication of a pure ferromagnetic behavior of the compound was found from the measured data. The relation of Θ_p to the magnetic transition temperature of (3.4 ± 0.3) K, as measured in an induction field of 2 mT, supports the assumption of spin-glass behavior [24]. In comparison to the low-field measurement (2 mT), the measurement in an induction field of 5 T shows a distinct broadening of the magnetic phase transition. This confirms that Rb$_4$O$_6$ is a magnetically frustrated system.

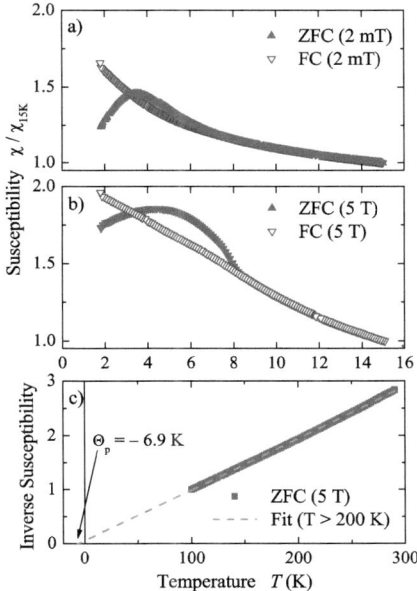

Figure 3.1: Magnetization data for rubidium sesquioxide. The low temperature behavior of the magnetic susceptibility is shown in (a) and (b) for induction fields of 2 mT and 5 T, respectively. The high temperature behavior of the inverse susceptibility for an expanded temperature scale is shown in (c). (All values are normalized by the value of the susceptibility at 15 K).

In contrast to recent suggestions [7], the present measurements of the magnetic properties do not confirm that Rb$_4$O$_6$ is a half-metallic ferromagnet. Instead, our data indicate that Rb$_4$O$_6$ may belong to the class of materials exhibiting thermally acquirable magnetically bistable states [25]. In any case, our measurements provide strong evidence that rubidium sesquioxide Rb$_4$O$_6$ is a magnetically frustrated system that exhibits spin-glass-like behavior, in a magnetic field. This peculiar behavior may be caused by a random charge ordering of the three charged dioxygen molecules in each formula unit. The incompatibility between the magnetization data presented here and the predicted half-metallic ferromagnetism seems to be due to an inadequate treatment of the molecular states in the local spin density approximation, which was used in the calculations by Attema et al. [7]. Moreover, the LSDA is inappropriate for addressing these particular types of charged dioxygen molecules and their magnetic ordering. It is worthwhile to note that LSDA calculations on rubidium hyperoxide RbO_2 predict this compound to be a half-metallic ferromagnet as well [26]. RbO_2 (a pale yellow powder [27]), however, is known from experiments to be an insulating antiferromagnet with a Néel temperature of approximately 15 K [28]. This again indicates that the LSDA calculations underestimate the localized bonds between the oxygen atoms as well as their magnetic states. The failure of LSDA in describing the electronic structure correctly cannot be overcome by using on-site electron-electron correlations in the form of LDA+U. Calculations with varying U up to 13.6 eV are able to explain an antiferromagnetic but still metallic, and not an insulating, state for RbO_2. Other compounds with open p-shells and molecular like valence states may have similar properties, so that performing LSDA calculations on these compounds may also result in the prediction of unrealistic magnetic ground states.

Although Rb_4O_6 does not exhibit half-metallic ferromagnetism, the fact that this alkali sesquioxide shows a complicated magnetic ordering, compared to other alkali oxides, is an exceptional phenomenon that is deserving further investigation. Follow-up experimental as well as more detailed theoretical studies must be performed to fully explain and understand the magnetic state of the charged oxygen molecules in Rb_4O_6 and in the related isoelectronic, isostructural sesquioxide Cs_4O_6. During submission of this work, Attema et al. [29] have reported about half-metallicity in other sesquioxides. The authors proposed for Cs_4O_6, among others, a half-metallic state with a Curie temperature of 350 K. However, preliminary magnetic measurements on that compound did not confirm the predicted magnetic state [26]. Therefore, it can be concluded that the mixed oxides as proposed in Ref.[29] will also not show the predicted half-metallicity.

Acknowledgements

We are grateful for the fruitful discussions with M. Jourdan, G. Jakob and F. Emmerling.

4 Challenge of magnetism in strongly correlated open-shell *2p* systems

The text of this chapter is identical with the following publication:
J. Winterlik, G. H. Fecher, C. A. Jenkins, C. Felser, C. Mühle, K. Doll, M. Jansen, L. M. Sandratskii, and J. Kübler
Phys. Rev. Lett. **102**, 016401 (2009).

Abstract

We report on theoretical investigations of the exotic magnetism in rubidium sesquioxide Rb_4O_6, a model correlated system with an open $2p$ shell. Experimental investigations indicated that Rb_4O_6 is a magnetically frustrated insulator. The frustration is explained here by electronic structure calculations that incorporate the correlation between the oxygen $2p$ electrons and deal with the mixed-valent oxygen. This leads to a physical picture where the symmetry is reduced because one third of the oxygen in Rb_4O_6 is nonmagnetic while the remaining two thirds assemble in antiferromagnetic arrangements. A degenerate, insulating ground state with a large number of frustrated noncollinear magnetic configurations is confidently deduced from the theoretical point of view. These findings demonstrate in general the importance of electron-electron correlations in open shell p-electron systems.

Solid oxygen is the paradigm for p-electron-based magnetic ordering in compounds and is, together with NO, the only molecular crystal that carries a magnetic moment. It displays an antiferromagnetic (AFM) transition at a Néel temperature of 24 K [30], and with moderate pressure, planes in the AFM solid couple ferromagnetically [31]. With higher pressure the molecular crystal becomes metallic [32] and then superconducting [33].

The alkali-metal oxides constitute an intermediate step on the way from molecular oxygen to the ionic transition-metal oxides. The alkali-metal atoms transfer their electrons to the oxygen molecules resulting in ionic crystals with dioxygen anions that retain the molecular properties of solid oxygen [34]. In particular, the heavy alkali metals form compounds with oxidation states that range from metallic suboxides [35] to the ozonides [36]. In this Letter it will be shown that the alkali sesquioxides - Rb_4O_6 and Cs_4O_6 - are of principal importance. Since these compounds are open-shell $2p$ systems within a solid, quantum chemistry and condensed matter meet in an intriguing fashion.

The black color of Rb_4O_6 suggests that it has an unconventional electronic structure, since comparable alkali hyperoxides or peroxides are pale yellow or white. Rb_4O_6 is an insulator with a resistivity of approximately 0.04 MΩ·m [10]. The sesquioxide contains three dioxygen anions with two possible valencies: the closed-shell peroxide anion O_2^{2-} and the open-shell hyperoxide anion O_2^-. The hyperoxide corresponds to a charged oxygen molecule (radical) and localizes its single unpaired electron in an antibonding π^* orbital. This causes the rare phenomenon of anionogenic magnetic order. Anionogenic magnetism is also observed, for example, in rubidium hyperoxide RbO_2, an insulating antiferromagnet with a Néel temperature of 15 K [28]. Electronic structure calculations have been used to predict *half-metallic ferromagnetism* in canonical open-shell $2p$ systems such as nanographene [37] or hole-doped MgO [38]. Similar predictions were made for Rb_4O_6 [7], but the underlying calculations failed to give an accurate description of its insulating and magnetic properties.

Rubidium sesquioxide crystallizes in the Pu_2C_3 structure type. Inelastic neutron scattering studies confirmed the simultaneous presence of O_2^- and O_2^{2-} anions [22]. Magnetization measurements [8] indicate a complicated electronic structure with a magnetic transition at approximately 3.4 K. An effective magnetic moment of $m = 1.83$ μ_B per hyperoxide anion was deduced from a Curie-Weiss fit. Thermal irreversibilities were observed between zero-field-cooled (ZFC) and field-cooled (FC) magnetization measurements. These phenomena are specific to spin glasses and related random magnetic systems [24, 39]. Further experiments indicated that the magnetization of Rb_4O_6 shows a dynamic time dependent behavior below the magnetic transition [10]. The magnetization as a function of time follows an exponential law with a relaxation time of $\tau = (1852 \pm 30)$ s. Magnetic and geometric frustration are observed frequently in d-

4. Challenge of magnetism in strongly correlated open-shell 2p systems

electron systems such as the spinel LiV_2O_4 [40] or the cubic vanadates [41]. In Rb_4O_6, however, the magnetic moment is carried by the p electrons of the anionic hyperoxide molecules. Here we show for the first time that open-shell $2p$ compounds exhibit magnetic frustration and behave like other correlated d- or f-electron systems.

For a detailed analysis of the electronic and magnetic structures of Rb_4O_6, its crystal structure must be understood. Figure 4.1(a) shows a body-centered cubic unit cell of Rb_4O_6. It contains 12 dioxygen anions that are distinguished by their valency and their alignment along the principal axes. The experimental bond lengths for the hyperoxide and the peroxide anions are approximately $0.14a$ and $0.17a$, respectively. The cubic lattice parameter is approximately $a \approx (9.2 - 9.3)$Å depending on the temperature [42, 43]. The next-nearest-neighbor environment of one hyperoxide anion is highlighted in Figure 4.1(b). The arrows denote the total molecular moment in an arrangement justified below. An AFM configuration that satisfies all interactions between hyperoxide anions is impossible; the order is frustrated.

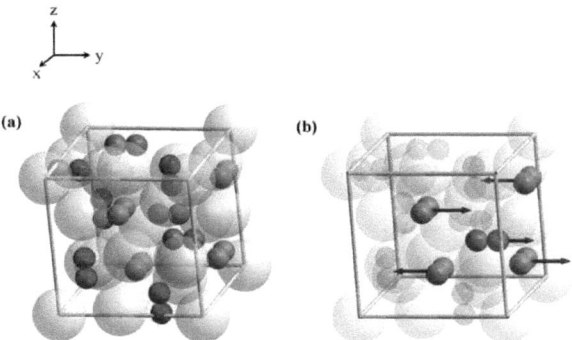

Figure 4.1: The pseudo-body-centered cubic cell of Rb_4O_6 is depicted in (a). Differently oriented dioxygen anions are drawn with different colors so that the alignment along the axes can be distinguished. For clarity, the Rb atoms are gray and transparent. The nonmagnetic peroxide anions (red) are assumed to be aligned along the z axis. The next-nearest-neighbor environment of one molecule consisting of hyperoxide anions is highlighted in (b). The vectors represent the total magnetic moment of an anion. The order is that found in the calculations.

Electronic structure calculations of Rb_4O_6 were performed using the quantum chemical CRYSTAL code [44] (details of the calculations are summarized in Ref. [45]). The implementation of the *exact exchange* Becke three-parameter Lee-Yang-Parr (B3LYP) hybrid functional [46] makes it useful for the theoretical treatment of molecular systems with a high degree of electron localization. A full optimization of the fractional coordinates was performed assuming that the dioxygen anions remain ferromagnetically ordered. A metallic ground state was obtained with space group $I\bar{4}3d$ where all dioxygen anions are symmetrically equivalent. The structural optimization resulted in a structure with lower symmetry (space group $I\bar{4}2d$) and different bond lengths of the peroxide and hyperoxide anions. Using the exact exchange hybrid functional, it converged to an insulating ground state as shown in Figure 4.1(a), where *all* peroxide anions are aligned along a single axis. The optimized bond lengths of $0.146a$ for the hyperoxides and $0.167a$ for the peroxides are in good agreement with the experimental values. The observed magnetic frustration cannot be calculated using the CRYSTAL code. Pure local density approximation (LDA) calculations result, however, in a metallic state [7], which is in disagreement with the experiments [10]. To overcome these difficulties, the augmented spherical wave (ASW) method [47] together with LDA plus the multiorbital mean-field Hubbard model LDA+U [48] was used to explain both the magnetic frustration and the insulating state (for details see Ref. [45]). The LDA+U method treats electron-electron correlations in narrow bands quite accurately in a mean-field sense. The Hubbard parameter U was implemented into the ASW code allowing for unrestricted noncollinear moment arrangements [49]. In the double counting correction scheme used here, the results depend only on the difference between U and the exchange parameter J. For simplicity, this difference is denoted by U in the following. A meaningful choice of U is not obvious *a priori*, but a discussion of possible values for oxygen in Ref. [50] and the experimentally determined large resistivity [10] suggest that values around 6 eV are a good choice.

The ASW LDA+U calculations converged only if the different bond lengths of the hyperoxide and peroxide pairs are accounted for. This is in agreement with the results of the cell optimization by CRYSTAL. The peroxide anions are chosen to be aligned along a single axis and found to be nonmagnetic. The hyperoxide anions are aligned along the orthogonal axes and couple antiferromagnetically with a magnetic moment of approximately 1 μ_B per pair [Fig. 4.1 (b)]. This accumulated moment does not depend on the choice of the parameter U. The individual contributions to the total moment, however, diverge from the original value of 0.5 μ_B as U increases. The symmetry is thus further reduced. The choice of axis for the peroxide anions is symmetrically equivalent, so it is assumed that the system spontaneously chooses an orientation for the peroxide anions that leads to antiferromagnetic order of the hyperoxide anions aligned along the perpendicular directions. Thus, the antiferromagnetism is degenerate and the peroxide anion is chosen to lie along the z-axis for all the following calculations.

Figures 4.2(a) and (b) show sections of the spin-resolved density of states in the vicinity of the Fermi energy for antiferromagnetic configurations using $U = 4.1$ eV and $U = 8.2$ eV. For these values, the gap that separates the occupied from the empty states is

4. Challenge of magnetism in strongly correlated open-shell 2p systems

$E_G \approx 0.1$ eV in (a) and $E_G = 0.8$ eV in (b). An intermediate value for the effective Hubbard parameter of $U = 6.8$ eV results in $E_G = 0.5$ eV (not shown). In all cases, a complete charge separation is obtained for the peroxide and hyperoxide anions. The magnetic moments of the peroxide anion are strictly zero. The highest occupied states are localized at the nonmagnetic peroxide anions. The s density originates from the Rb ions and supplies the bonding and exchange paths.

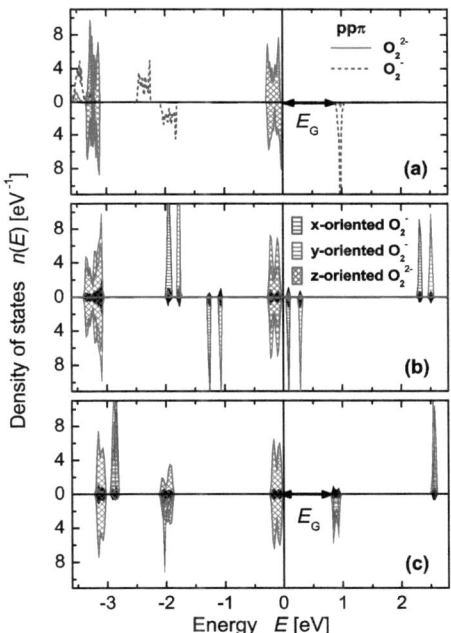

Figure 4.2: The density of states for Rb_4O_6 obtained by CRYSTAL with the exact exchange hybrid functional for a ferromagnetic configuration of the O_2^- anions is shown in (a). (b) and (c) show the sublattice density of states obtained by ASW with LDA+U for an antiferromagnetic setup [$U = 4.1$ eV (b) and $U = 8.1$ eV (c)]. Red shading is used for the nonmagnetic O_2^{2-} anions, blue and cyan for the magnetic O_2^- anions. Black shading indicates Rb s states. The upper halves are spin-up; the lower halves spin-down. The energy origin is chosen to be the Fermi energy. The value of the energy gap, E_G, that separates the highest occupied states from the lowest unoccupied states is $E_G \approx 0.9$ eV in (a), 0.1 eV in (b), and 0.8 eV in (c).

For comparison, Figure 4.2(a) exhibits the density of states in the ferromagnetic configuration obtained by CRYSTAL. The electronic structure is obviously insulating. The gap emerges from the splitting of the O_2^- and O_2^{2-} $pp\pi$ orbitals. All $pp\sigma$ states are too far below the Fermi energy to affect the electronic character of Rb_4O_6. Starting with a ferromagnetic configuration in the ASW calculations with values of U below 4.1 eV, the electronic structure became half-metallic ferromagnetic. At $U = 4.1$ eV a magnetic moment of exactly 4 μ_B per unit cell was obtained. The peroxide anions become weakly magnetic with an induced moment of 0.1 μ_B, decreasing to 0.05 μ_B for the largest value of U. The ferromagnet became insulating for $U = 6.8$ eV and 8.2 eV.

The ferromagnet is now excluded from our consideration by using the total energy criterion. For $U = 4.1$ eV, it is 87 meV per unit cell higher for the ferromagnet relative to the antiferromagnetic state. For even larger values of U, the stability of the antiferromagnetic state decreases slightly to approximately 56 meV per unit cell for $U = 8.1$ eV. It is thus the electron correlation that leads to an antiferromagnetic ground state, although it cannot be definitively asserted that the true ground state is exactly as depicted in Figure 4.1(b).

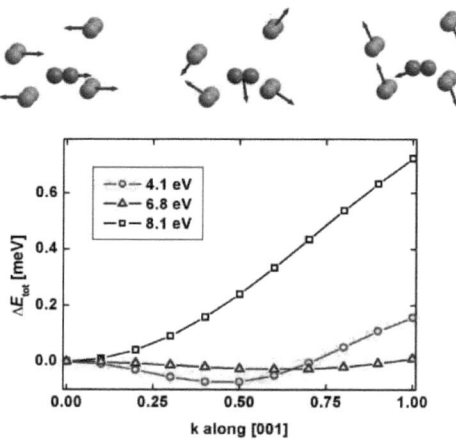

Figure 4.3: The line plot shows the spiral energies of antiferromagnetic Rb_4O_6 as functions of the spiral vector **k** in units of $2\pi/a$ along the $[0,0,1]$-direction for $U = 4.1$ eV, 6.8 eV, and 8.1 eV. The corresponding magnetic configurations are sketched for $k = 0$, $\mathbf{k} = (0,0,0.5)$, and $\mathbf{k} = (0,0,1)$. The energy is given per primitive cell.

It is nontrivial to find the true ground state out of the large number of possible noncollinear spin orientations. Spiral modulations of the antiferromagnet enable an improved search [49, 51], provided that spin-orbit interaction can be neglected (as justified here [45]). A magnetic spiral is defined by the size of the magnetic moment, an angle of tilt θ, and a wave vector $\mathbf{k}$ that can be chosen in the Brillouin zone of the crystal. For convenience, a tilt angle of $\theta = 90°$ is used. Such a spin spiral changes the polar angle of the moment of the ion with basis vector $\mathbf{x}$ by $\phi = 2(\mathbf{k} \cdot \mathbf{x})$.

Spin spirals with various values of the wave vector $\mathbf{k}$ are thus superimposed onto the antiferromagnetic state. Assuming the force theorem to be a good approximation, the total energy changes ΔE_{tot} for each value of $\mathbf{k}$ are calculated. The resulting values of ΔE_{tot} are extremely small for $\mathbf{k}$ along the [001] direction. The energy changes are shown in Figure 4.3 for $U = 4.1$ eV, 6.8 eV, and 8.1 eV. The spiral energies for any direction in the BZ other than [001] are larger by three to four orders of magnitude because the parallel coupling within the hyperoxide anion is extremely strong and the moments remain parallel only for $\mathbf{k}$ along [001].

The frustration of the magnetic moments is clearly seen in the antiferromagnetic case [see Figures 4.1(b) and 4.3] where the alternating ferromagnetic and antiferromagnetic bonds are obvious). The magnetic moments along the next-nearest neighbor connections are perpendicular for $\mathbf{k} = (0, 0, 1)$ at the zone boundary and are thus no longer frustrated. The flat and slightly negative portions of ΔE_{tot} for $U = 4.1$ eV and 6.8 eV may be assumed to serve as a reservoir of states to release the frustration, although it should be noticed that the part of k space available along [001] is extremely restricted. A theoretical description of the electronic and magnetic state of the model alkali-oxide system Rb_4O_6 that is compatible with experimental results [10] is thus only possible by taking into account the electron-electron correlations, a complex magnetic arrangement of the spins, and the reduction of the symmetry. In particular, the existence of nonmagnetic peroxide anions is essential and results in a peculiar electronic texture in real and reciprocal space.

In summary, it has been shown that the proposed half-metallic ferromagnetism [7] in Rb_4O_6 is unrealistic, and instead an insulating ground state is obtained. The magnetic order is found to be quite exotic in so far as one third of the oxygen becomes nonmagnetic while the remaining two thirds assemble in a frustrated antiferromagnetic configuration. The magnetic frustration and the spin spiral state in Rb_4O_6 as well as its multidegenerate ground state were demonstrated using noncollinear spin calculations. The calculations explain both the strong time dependence of the magnetization and the pronounced differences between the ZFC and FC measurements that are characteristic of a frustrated magnetic state [10]. The frustration is of geometric origin and caused by the peculiar symmetry of Rb_4O_6. The importance of electronic correlation in a $2p$ compound was demonstrated here for Rb_4O_6. Open-shell $2p$ systems thus behave like $3d$ or $4f$ systems. Strong correlation is also expected to be important in other p-electron systems. In this Letter it was shown that Rb_4O_6 serves as a model for other open-shell systems that are based on p-electrons.

Acknowledgements

This work was funded by the DFG in the Collaborative Research Center *Condensed Matter Systems with Variable Many-Body Interactions* (TRR 49). The authors are grateful for the fruitful discussions with W. Pickett, M. Jourdan, G. Jakob and H. von Löhneysen.

5 Exotic magnetism in the alkali sesquoxides Rb_4O_6 and Cs_4O_6

The text of this chapter is identical with the following manuscript, which is submitted for publication:
J. Winterlik, G. H. Fecher, C. A. Jenkins, C. Felser, J. Kübler, C. Mühle, M. Jansen, T. Palasyuk, I. Trojan, S. Medvedev, M. I. Eremets and F. Emmerling
Accepted for publication in Phys. Rev. B (2009).

Abstract

Among the various alkali oxides the sesquioxides Rb_4O_6 and Cs_4O_6 are of special interest. Electronic structure calculations using the local spin-density approximation predicted that Rb_4O_6 should be a half-metallic ferromagnet, which was later contradicted when an experimental investigation of the temperature dependent magnetization of Rb_4O_6 showed a low-temperature magnetic transition and differences between zero-field-cooled (ZFC) and field-cooled (FC) measurements. Such behavior is known from spin glasses and frustrated systems. Rb_4O_6 and Cs_4O_6 comprise two different types of dioxygen anions, the hyperoxide and the peroxide anions. The nonmagnetic peroxide anions do not contain unpaired electrons while the hyperoxide anions contain unpaired electrons in antibonding π^*-orbitals. High electron localization (narrow bands) suggests that electronic correlations are of major importance in these open shell p-electron systems. Correlations and charge ordering due to the mixed valency render p-electron-based anionogenic magnetic order possible in the sesquioxides. In this work we present an experimental comparison of Rb_4O_6 and the related Cs_4O_6. The crystal structures are verified using powder x-ray diffraction. The mixed valency of both compounds is confirmed using Raman spectroscopy, and time-dependent magnetization experiments indicate that both compounds show magnetic frustration, a feature only previously known from d- and f-electron systems.

Introduction

Magnetism arising from d- and f-shells has drawn the bulk of research attention but p-electron based magnetic order is a rare and fascinating topic that presents the added challenge of molecular and not just atomic ordering. The majority of main group molecules are nonmagnetic. Few exceptions are found, e.g. in NO, NO_2 and O_2. Molecular oxygen contains two single electrons in degenerate antibonding π^*-orbitals, which can order magnetically in a solid crystal. Solid oxygen shows a large variety of magnetic phenomena ranging from antiferromagnetism to superconductivity [30–33, 52].

What applies to molecular oxygen applies equally to charged oxygen molecules with unpaired electrons. Dioxygen anions are principally found in alkali and alkaline earth oxides, which represent excellent model systems because of their supposedly simple electron configurations. The hyperoxide anion O_2^- corresponds to "charged oxygen". Since it still contains one unpaired electron, magetic order is enabled for hyperoxides. KO_2, RbO_2, and CsO_2 are known to exhibit antiferromagnetic ordering below their respective Néel temperatures of 7 K, 15 K, and 9.6 K, respectively [28, 53].

Among the alkali oxides, the sesquioxides are of special interest. In contrast to related compounds, which are white, yellow, or orange, the sesquioxides Rb_4O_6 and Cs_4O_6 are black. Furthermore, a formula unit AM_4O_6 (AM = alkali metals Rb or Cs) contains two different types of dioxygen anions: one closed-shell nonmagnetic peroxide anion and two of the aforementioned hyperoxide anions. The structural formula of the sesquioxides is thus accurately represented as $(AM^+)_4(O_2^-)_2(O_2^{2-})$ [20]. The mixed valency enables complicated magnetic structures in the sesquioxides. The first descriptions of the crystal structures of the sesquioxides were published in 1939 in the pioneering works of Helms and Klemm [21, 54, 55]. Both compounds belong to the Pu_2C_3 structure type and to space group $I\bar{4}3d$. This Pu_2C_3 structure type is known from the noncentrosymmetric rare earth metal sesquicarbide superconductors such as Y_2C_3 with a maximum critical temperature of $T_c = 18$ K [56]. For Cs_4O_6, the literature runs out after 1939 because of the extremely challenging synthesis and sensitivity to air. In Rb_4O_6, the presence of both peroxide and hyperoxide anions was verified by neutron scattering [22], and electronic structure calculations using the local spin density approximation (LSDA) were performed to explain the exceptional black color [7]. These same calculations predicted a half-metallic ferromagnetic ground state but were contradicted by later experiments in which a transition was found to occur at approximately 3.4 K in the temperature dependent magnetization [8]. Differences between ZFC and FC measurements indicated that Rb_4O_6 behaves like a frustrated system or a spin glass. Recently published electronic structure calculations are consistent with the experimental findings [9]. It was shown that for an accurate theoretical treatment of highly localized systems such

5. Exotic magnetism in the alkali sesquoxides Rb_4O_6 and Cs_4O_6

as Rb_4O_6 and Cs_4O_6, a symmetry reduction, exact exchange and electron-electron correlations have to be considered in the calculations. This can be generalized to any other open shell system that is based on p-electrons. Accounting for these features the calculations result in an insulating ground state. Further calculations using the spin spiral method show that Rb_4O_6 exhibits spin spiral behavior in a certain crystal direction. The energy changes are extremely small along this direction indicating a multidegenerate ground state [9]. In this work we present the routes of synthesis for Rb_4O_6 and Cs_4O_6 and the structural verification using powder x-ray diffraction (XRD). Raman spectroscopic measurements confirm the mixed valency of the used Rb_4O_6 and Cs_4O_6 samples. Magnetization experiments are shown that indicate a dynamic time dependent magnetism of Rb_4O_6. Furthermore we present a comprehensive experimental study of Cs_4O_6, which has not been extensively investigated due to the difficulty of sample preparation. Magnetization measurements dependent on temperature, magnetic field, and time provide evidence that Cs_4O_6 shows a similar behavior as Rb_4O_6. According to experimental and theoretical investigations [9], both sesquioxides are mixed valent highly correlated systems exhibiting p-electron based magnetic frustration. These seemingly simple compounds can serve as model systems for any other open-shell systems that are based on p-electrons such as hole-doped MgO [38] or nanographene [37].

Structure

Figure 5.1 depicts a body-centered cubic unit cell of AM_4O_6. The 24 oxygen atoms in the cubic cell form 12 molecules that can be distinguished both by their valency and by their alignment along the principal axes. We assumed the nonmagnetic peroxide anions to be oriented along the z-axis, whereas the hyperoxides are oriented along the x- and y-axes. The experimental bond lengths for the hyperoxide and the peroxide anions are 0.144 a and 0.165 a, respectively [42, 43]. The cubic lattice parameters are found in Section 5.

Synthesis

The precursors rubidium and cesium oxide AM_2O as well as rubidium and cesium hyperoxide AMO_2 were prepared from elemental sources. For AM_2O, liquid rubidium/cesium, purified by distillation, was reacted with a stochiometric amount of dry oxygen in an evacuated glass tube followed by heating at 473 K for two weeks under argon atmosphere [27, 57]. The samples were subsequently ground under argon and the entire cycle was repeated five times. A slight excess of rubidium/cesium was distilled at 573 K in vacuum, resulting in a pale green powder of Rb_2O and an orange powder of Cs_2O. AMO_2 were prepared through the reaction of liquid rubidium/cesium and an excess of dry oxygen using the same method as described for AM_2O, resulting in yellow powders of RbO_2 and CsO_2, respectively.

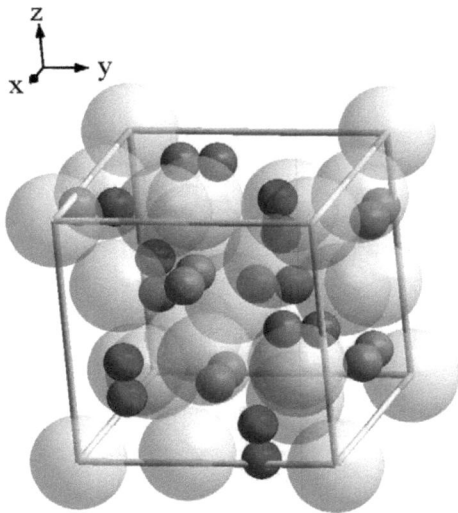

Figure 5.1: Pseudo-body-centered cubic cell of the alkali sesquioxides AM_4O_6 (AM = Rb, Cs). Differently oriented dioxygen anions are drawn with different colors so that the alignment along the axes can be distinguished. For clarity, the Rb/Cs atoms are gray and transparent.

Rb_4O_6 was obtained by a solid state reaction of 400 mg RbO_2 and 160 mg Rb_2O in a molecular ratio of 4 : 1 in a glass tube sealed under argon at 453 K for 24 hours. Cs_4O_6 was obtained in the same way using the amounts of 468 mg CsO_2 and 280 mg Cs_2O and annealing at 473 K for 24 hours. For the reaction, a slight excess of Cs_2O was used to compensate for small amounts of cesium peroxide, which is always contained as an imurity in Cs_2O. The very air sensitive products were ground and the reaction was repeated until pure phases were obtained (black powders). For the magnetization measurements, Rb_4O_6 and Cs_4O_6 were sealed in a high-purity quartz tube (Suprasil glass) under helium atmosphere.

Structural characterization

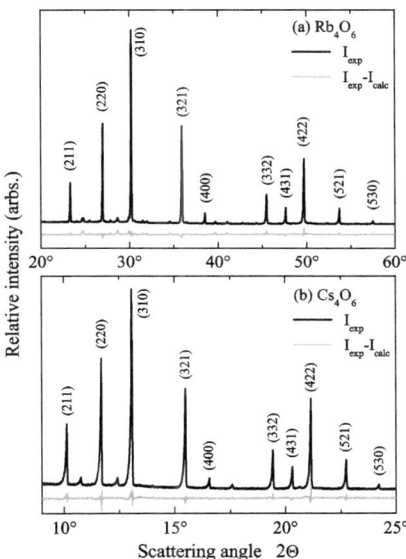

Figure 5.2: Powder x-ray diffraction patterns of Rb_4O_6 and Cs_4O_6 at 300 K (black). The gray curves show the differences between the observed data and the Rietveld refinements.

The crystal structures of the compounds were investigated using XRD. The measurements were carried out using a Bruker D8 diffractometer with Cu $K_{\alpha 1}$ radiation for Rb_4O_6 and Mo $K_{\alpha 1}$ for Cs_4O_6. The diffraction patterns are shown in Figure 10.5. The raw data (black) are compared to the difference between a calculated Rietveld refinement and the raw data (gray). The refinements yielded weighted profile R-values of $R_{wp} = 8.216$ for Rb_4O_6 and 6.388 for Cs_4O_6. Both compounds crystallize in the cubic structure $I\bar{4}3d$ (space group 220). The experimental lattice parameters as found from Rietveld refinements are 9.322649(74) Å for Rb_4O_6 and 9.84583(11) Å for Cs_4O_6. The atomic parameters for both compounds are shown in Table 5.1. The pattern of Rb_4O_6 indicates good phase purity. In the case of Cs_4O_6, additional signals were detected. The signals were identified to arise from the impurity CsO_2, which belongs to the cubic

Table 5.1: Atomic parameters for Rb_4O_6 and for Cs_4O_6

Atom	Site	x	y	z
Rb	16c	1.054696(50)	1.054696(50)	1.054696(50)
O	24d	1.20206(36)	0	3/4
Cs	16c	0.946544(45)	0.946544(45)	0.946544(45)
O	24d	0.55065(49)	0	3/4

space group $Fm\overline{3}m$ and has a lattice parameter of $a = 6.55296(37)$ Å. The impurity phase was included in the refinement of Cs_4O_6. An impurity content of approximately 7.23% of CsO_2 in Cs_4O_6 was derived from the refinement.

Raman Spectroscopy

The presence of peroxide and hyperoxide anions in Rb_4O_6 and Cs_4O_6 was verfied using Raman spectroscopy. The measurements were performed using a diamond anvil cell (DAC) in order to prevent sample decomposition from contact with air and moisture. The samples were loaded in a dry box under dry nitrogen atmosphere. The samples were confined within a cylindrical hole of 100 microns diameter and 50 microns height drilled in a Re gasket. We used synthetic type-IIa diamond anvils, which have only traces of impurities (<1 ppm) and very low intrinsic luminescence. Raman spectra were recorded with a single 460-mm-focal-length imaging spectrometer (Jobin Yvon HR 460) equipped with 900 and 150 grooves/mm gratings, giving a resolution of 15 cm^{-1}, notch-filter (Kaiser Optics), liquid nitrogen cooled charge-coupled device (CCD (Roper Scientific)). Scattering calibration was done using Ne lines with an uncertainty of ±1 cm^{-1}. The He-Ne laser of Melles Griot with the wavelength of 632.817 nm was used for excitation of the sample. The probing area was a spot with 5 microns diameter. Figure 5.3 shows the Raman spectrum of Rb_4O_6 in a range from 700-1250 cm^{-1}. Two peaks are found at Raman shifts of 795 cm^{-1} and 1153 cm^{-1}, respectively. The signal at 795 cm^{-1} corresponds to the stretching vibration of the peroxide anions and is in good agreement with the literature value of 782 cm^{-1} for Rb_2O_2 [58]. The peak at 1153 cm^{-1} is assigned to the corresponding vibration of the hyperoxide anions and comparable to the literature value of 1140 cm^{-1} for RbO_2 [59]. In the Raman spectrum for Cs_4O_6, which was recorded as described above, signals of 738 cm^{-1} and 1128 cm^{-1} were recorded. The simultaneous presence of both dioxygen anion types and thus the mixed valency is proven for both Rb_4O_6 and Cs_4O_6.

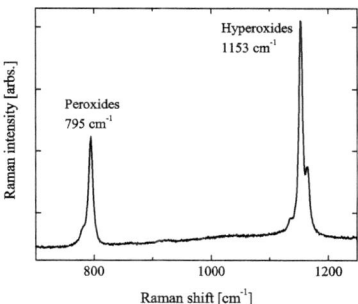

Figure 5.3: Raman spectrum of Rb_4O_6. The peak at 795 cm^{-1} corresponds to the stretching vibration of the peroxide anions and the peak at 1153 cm^{-1} to the stretching vibration of the hyperoxide anions.

Magnetization

The magnetic properties of Rb_4O_6 and Cs_4O_6 were investigated using a superconducting quantum interference device (SQUID, Quantum Design MPMS-XL5). Samples of approximately 100 mg, fused in Suprasil tubes under helium atmosphere, were used for the analysis. In the cases of temperature dependent and time dependent magnetometry we performed under zero-field-cooled (ZFC) and field-cooled (FC) measurements. For the ZFC conditions, the samples were first cooled to a temperature of 1.8 K without applying a magnetic field. After applying an induction field $\mu_0 H$, the magnetization was recorded as a function of temperature or time. For the temperature dependent measurements, the magnetization was recorded directly afterward in the same field upon lowering the temperature down to 1.8 K again (FC).

Temperature dependent magnetization experiments were discussed in an earlier work [8] and indicate that Rb_4O_6 behaves like a frustrated system. We have analyzed the related Cs_4O_6 using an identical experimental setup. Figures 5.4(a) and (b) display the temperature dependent magnetization of Cs_4O_6 at magnetic induction fields $\mu_0 H$ of 2 mT and 5 T, respectively. The measurements were carried out under ZFC conditions and FC conditions. In the 2 mT ZFC measurement, a magnetic transition is found to occur at 3.2 ±0.2 K. In the 5 T ZFC curve, this transition shows a distinct broadening and is shifted to a higher temperature. Thermal irreversibilities between ZFC and FC measurements are observed in both magnetic fields, a behavior known from frustrated systems and spin glasses. Cs_4O_6 exhibits similar magnetic properties as Rb_4O_6. Figure 5.4(c) shows the normalized inverse susceptibility of Cs_4O_6 in a temperature range of 100-300 K at a magnetic field of 5 T. A Curie-Weiss fit was performed in the region above 200 K and yields a negative paramagnetic transition temperature Θ_D=-4.5 K, which indicates dominance of antiferromagnetic interactions. From the high temperature data, an effective magnetic moment of $m = 2.01\mu_B$ per hyperoxide anion can be deduced applying the Curie-Weiss law based on molecular field theory (MFT). This value is in modest agreement with 1.73 μ_B as expected from MFT using the spin-only approximation.

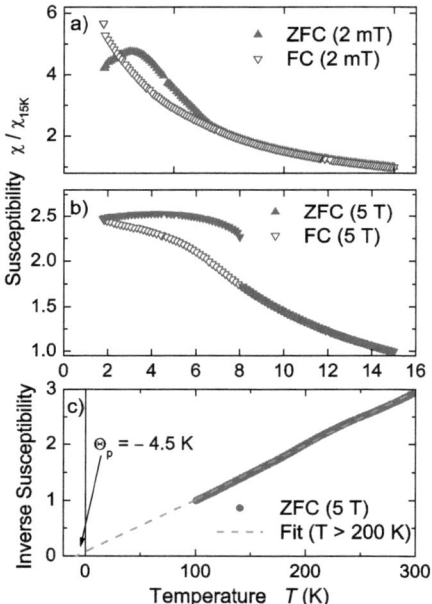

Figure 5.4: Shown is the temperature dependent magnetic susceptibility of Cs_4O_6. The low-temperature behavior is shown in (a) and (b) for induction fields of 2 mT and 5 T, respectively. The high-temperature behavior is shown in (c). The dashed line represents a Curie-Weiss-Fit in the temperature region above 200 K. (All values are normalized by the value of the susceptibility at 2 K.)

We also performed field dependent magnetization experiments. Figure 5.4(a) shows the field dependent magnetization of Rb_4O_6 at 2 K, well below the magnetic transition temperature of 3.4 K. It is clear that the magnetization does not show hysteresis [see inset (i)]. The shape of the magnetization curve is best modelled by a paramagnetic loop correction to a Langevin function $L(H)$ given by $M(H) = \chi_{lin}H + M_0 L(H)$, where χ_{lin} is the field independent paramagnetic susceptibility and M_0 the saturation moment.

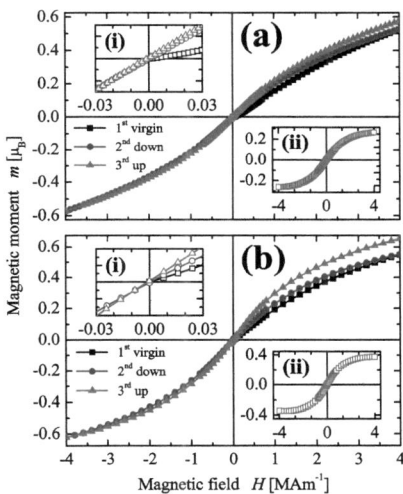

Figure 5.5: Field-dependent magnetization of Rb_4O_6 and Cs_4O_6 at T = 2 K. Shown are three magnetization cycles as a function of applied H: virgin $(0 \to +H_{max})$, down $(+H_{max} \to -H_{max})$, and up $(-H_{max} \to +H_{max})$. The insets (i) show in detail the magnetization close to the origin. The insets (ii) show the remaining Langevin functions after subtracting the linear paramagnetic correction.

In inset (ii), the linear contribution was subtracted from the total magnetization. The remaining Langevin function saturates initially with a magnetic moment of 0.25 μ_B per Rb_4O_6 formula unit and increases in successive cycles. At given magnetic fields, the up and down curves exhibit differences in the measured magnetization, indicating that the magnetization changes with time in a manner consistent with the known relaxation behavior of frustrated systems [60]. Figure 5.4(b) shows the corresponding field dependent magnetization for Cs_4O_6, qualitatively similar to Rb_4O_6. Applying the model of the paramagnetic loop correction to a Langevin function as described above a saturation magnetic moment of 0.37 μ_B per Cs_4O_6 formula unit is obtained as seen in Figure 5.4(b) inset (ii). While the total magnetic moments of Rb_4O_6 and Cs_4O_6 are quite similar at given magnetic fields, the paramagnetic contributions exhibit a large difference. This indicates that the time dependence and the dynamics of the compounds are differing.

Time dependent measurements reveal more about these dynamics. Figures 5.5(a) and (b) show the time dependent variations of the magnetization up to 6400 s for both compounds. The measurements were performed under ZFC conditions at 2 K in induction fields of 1 T. The magnetic moment of Rb_4O_6 varies exponentially with a relaxation time of $\tau = (1852\pm30)$ s. This value is comparable to those of other frustrated systems. Similar curves were obtained for Rb_4O_6 using lower as well as higher induction fields of 30 mT and 5 T. The relaxation times were determined to be $\tau(30\text{ mT}) = (1170 \pm 30)$ s and $\tau(5\text{ T}) = (4340 \pm 20)$ s. The pronounced relaxation is another clear indication of the magnetic frustration in Rb_4O_6.

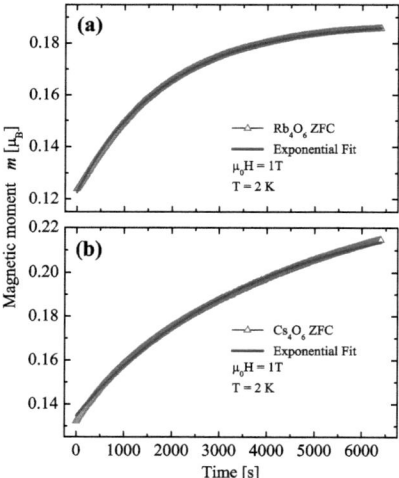

Figure 5.6: Time dependent magnetizations of Rb_4O_6 and Cs_4O_6 in induction fields of $\mu_0 H = 1$ T are shown. The solid lines are the result of exponential fits.

As in the case of Rb_4O_6, the magnetic moment of Cs_4O_6 follows exponential behavior. A relaxation time of $\tau = (3701\pm30)$ s is deduced from the exponential fit confirming that

Cs_4O_6 is also a magnetically frustrated $2p$-system although with different dynamics. The fitting curve does not follow the exponential behavior as exactly as for Rb_4O_6. This is most probably due to the fact that a complete saturation of the magnetization is not reached within the time span of 6400 s. A reason for these differences between the compounds cannot be given within the scope of these experiments, but electronic structure calculations for Cs_4O_6 may shed more light on the dynamics of this system.

Summary and Conclusions

It has been shown that the alkali sesquioxides Rb_4O_6 and Cs_4O_6 exhibit frustrated magnetic ordering based on anionogenic $2p$-electrons of the hyperoxide anions. Mixed valency was verified in both compounds using Raman spectroscopy. The strong time dependence of the magnetization and the pronounced differences between the ZFC and FC measurements support the previously assumed frustrated state of Rb_4O_6 [9]. Cs_4O_6 was found to show a very similar behavior. The experiments show that strong electronic correlations can also occur in presumably simple $2p$-compounds such as alkali oxides. The complex distribution of magnetic moments in the lattices leads to a symmetry reduction and causes a frustrated magnetic beahvior in both compounds. These results are of major importance since they confirm that open shell p-electrons can behave like d- or f-electrons.

Acknowledgements

This work was funded by the DFG in the Collaborative Research Center *Condensed Matter Systems with Variable Many-Body Interactions* (TRR 49). The authors are grateful for the fruitful discussions with W. Pickett, M. Jourdan, G. Jakob and H. von Löhneysen.

6 Electronic and structural properties of palladium-based Heusler superconductors

The text of this chapter is identical with the following publication:
J. Winterlik, G. H. Fecher, and C. Felser
Solid State Commun. **145**, 475 (2008).

Abstract

This work reports on superconductivity in the Heusler compounds Pd_2ZrAl and Pd_2HfAl. Magnetization and resistance measurements were carried out to verify their superconducting states. The compounds exhibit transition temperatures of 3.2 K (Zr) and 3.4 K (Hf). From their behavior in external magnetic fields, it was determined that both compounds are type II superconductors. Similar to the half-metallic ferromagnets, the superconducting Heusler compounds follow an electron counting scheme based on theoretical considerations. As found from *ab initio* calculations, the superconductivity can be explained by a valence instability at the L-point, that has been used as design criterion.

Introduction

Conventional superconductivity and ferromagnetism are two mutually exclusive phenomena, since conventional Cooper pairs consist of electrons with opposite spins, which build up a singlet spin state leading to s-wave superconductivity. Because ferromagnetism destroys singlet pairs, contact with ferromagnets suppresses conventional superconductivity. For this reason, it was believed that supercurrents could be induced in ferromagnetic materials only via unconventional triplet superconductors. Recently, Keizer et al. developed a device in which two conventional superconducting layers have been coupled by a layer of the half-metallic ferromagnetic (HMF) CrO_2 [61, 62]. The report of a spin triplet supercurrent through the CrO_2 layer proves that singlet Cooper pairs were converted to triplet Cooper pairs. This indicates that there is a potential for designing magnetization-controlled Josephson junctions, which consist of conventional superconductors and half-metallic ferromagnets, epitaxially grown on suitable substrate.

The class of Heusler compounds A_2BC that crystallizes in the cubic $L2_1$ structure, is well known for magnetic ordering. Up to now, many of the Heusler compounds have been reported to be half-metallic ferromagnets [62, 63] and several Co_2-based Heusler compounds have been successfully implemented as electrodes in magnetic tunnel junctions [64, 65]. In contrast to the half-metallic ferromagnets, only a few Heusler compounds, all of them with a rare earth metal at the B position, have been reported to be superconductors [66]. Until now, Pd_2YSn has been found to be the Heusler compound with the highest critical temperature (4.9 K) [67]. Coexistence of superconductivity and antiferromagnetism, demonstrating the manifoldness of the Heusler family, was reported for Pd_2YbSn [68] and Pd_2ErSn [69].

The occurrence of superconductivity in Heusler compounds brings up the task of building a bridge from magnetism to superconductivity. The results of Keizer et al. [61] are the motivation behind proposing a magnetization-controlled junction that is based only on Heusler compounds. For such a junction, a half-metallic ferromagnetic Heusler compound and a Heusler superconductor with matching lattice parameters are needed. Since a large variety of Heusler HMFs with a broad range of lattice parameters is already available, the remaining task is to find suitable Heusler superconductors for building such a device.

For a specific material design, it is necessary to have a well-established recipe for predicting the electronic as well as the magnetic properties of the particular class of materials. The Bardeen-Cooper-Schrieffer theory states that the transition temperature of a superconducting material, at given Debye frequency and Cooper-pairing interaction, increases exponentially with an increasing density of states $N(\epsilon_F)$ at the Fermi energy ϵ_F. This implies that, for a particular class of materials, a valence electron concentration that results in a maximum of $N(\epsilon_F)$ should yield a high transition temperature [70]. For elements, the dependence of T_c on the valence electron concentration is known as Matthias' rule [71]. Similarly, high critical temperatures are observed in A15 compounds at distinguished electron concentrations of about 4.6 and 6.4 electrons per atom [72]. According to L. van Hove, the electronic structure of a metal exhibits extrema at certain high symmetry points [14] that are referred to as van Hove singularities and leads to a peak in the density of states. The van Hove scenario provides an explanation for the unusually high transition temperatures of the A15 superconductors, which have been of interest up to the present with regard to their use in applications (mainly Nb$_3$Sn, with a critical temperature T$_c$ of 18 K, as a material for superconducting magnets) [73]. Following the van Hove scenario [15], superconductivity is predicted here for Heusler compounds with 27 electrons that exhibit valence instabilities close to ϵ_F, and in particular, for a saddle-point at the L-point.

Results and discussion

To find suitable compounds using the above given design criteria, the electronic structure of the Heusler compounds Pd$_2$ZrAl and Pd$_2$HfAl has been calculated *ab initio*. The electronic structure of the two compounds has been calculated using the full potential linearized augmented plane wave method as implemented in WIEN2k [74]. The exchange correlation functional was taken within the generalized gradient approximation [75]. A $25 \times 25 \times 25$ point mesh was used as the base for the integration. The convergence criteria were set to 10^{-5} Ry for energy and, simultaneously, $10^{-3}e^-$ for charge. The muffin tin radii were set to about 1.3 Å for all elements. The electronic structures presented below were calculated for the experimental as well as the optimized lattice parameters of each compound. The equilibrium lattice parameters from an equation of states fit [76] are slightly larger by approximately 1% than those obtained from experiment (see Table 6.1).

The calculated electronic structure of Pd$_2$HfAl is shown in Figure 6.1. The electronic structure of the corresponding compound Pd$_2$ZrAl is very similar. In the band structure as well as in the density of states, the hybridization gap - typical for Heusler compounds - that separates the low lying s-states from the d-bands is clearly visible at approximately 6 eV below the Fermi energy. This gap is slightly larger in Pd$_2$ZrAl, indicating that the lattice parameter for Pd$_2$ZrAl is larger than for Pd$_2$HfAl.

Most interesting, however, is the occurrence of a van Hove singularity just above the Fermi energy at the L-point (note that the Λ band, defining the saddle point close to ϵ_F, has only a weak positive 2nd derivative at L that is not resolved in Fig. 6.1(a). In addition, only one, rather steep, d-band crosses the Δ-direction. The flat bands forming the van Hove singularity at the L-point result in a maximum in the density of states.

Table 6.1: Energy of the valence band instabilities near ϵ_F. All energies are given in meV with respect to the Fermi energy ϵ_F. The lattice parameters are given in Å (the experimental and the optimized values for a are indicated by the subscripts $_{exp,opt}$, respectively).

Compound	a	$E(L)$	$E(\Gamma)$	$E(X)$
Pd$_2$ZrAl	6.390$_{exp}$	108	- 694	- 317
	6.467$_{opt}$	109	- 630	- 299
Pd$_2$HfAl	6.370$_{exp}$	172	- 643	- 287
	6.436$_{opt}$	173	- 585	- 272

This maximum, emerging from the states just above ϵ_F, is approximately 3.5 eV^{-1} for the Zr compound, which is approximately 0.5 eV^{-1} higher than the maximum for the Hf compound. In both compounds, the density is slightly lower directly at ϵ_F. From this point of view, the above given design criteria for a superconductor are fulfilled.

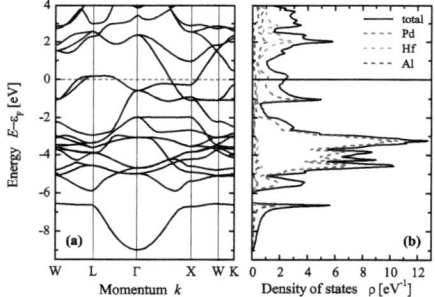

Figure 6.1: Electronic structure of Pd$_2$HfAl. The band structure is given in (a) and the density of states is given in (b), both are calculated for the optimised lattice parameter.

The energies of the states at the high symmetry points are summarized in Table 6.1. From Table 6.1, it may be seen that the energetical position of the van Hove singularity at the L-point is rather stable against variations in the lattice parameter, whereas some deviations are found in the energies of the states at the Γ and X-points when the values calculated for the experimental and optimized lattice parameters are compared.

Polycrystalline samples were prepared by repeated arc melting of a stoichiometric mixture of the constituents under an argon atmosphere at a pressure of 10^{-4} mbar. The resulting polycrystalline ingots were annealed afterwards for 7 days at 1073 K in an evacuated quartz tube. This procedure resulted in samples exhibiting the $L2_1$ struc-

6. Electronic and structural properties of palladium-based Heusler superconductors

ture. The crystal structure was determined by X-ray powder diffraction (XRD) using excitation by Cu K_α radiation. According to the XRD data, neither compound exhibited any impurities. The lattice parameters found from a Rietveld-refinement are given in Table 6.1.

The magnetic properties were investigated using a superconducting quantum interference device (Quantum Design, MPMS-XL-5) using small spherical pieces of approximately 20 mg to 120 mg of the samples. The transport properties were investigated by a physical property measurement system (Quantum Design, PPMS Model 6000). The resistance of samples with polished surfaces was measured using the four-point contact method.

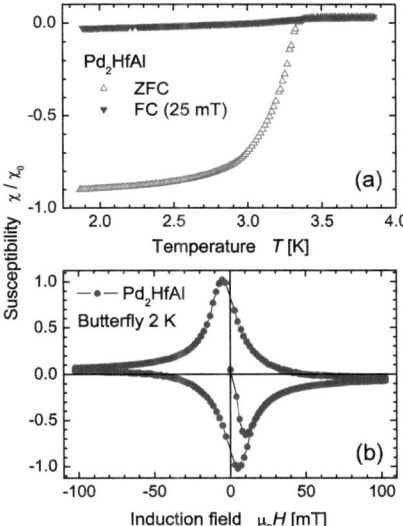

Figure 6.2: Magnetic measurements for Pd$_2$HfAl. (a) shows the temperature-dependent magnetization under ZFC and FC conditions. (b) shows the field-dependent magnetization (butterfly) curve at a temperature of 2 K.

The results of the magnetization measurements, plotted for Pd$_2$HfAl as an example, are given in Figure 6.2. The upper panel shows the temperature-dependent magnetization $M(T)$ of the sample in an external induction field of 25 mT for both zero-field cooled (ZFC) and field-cooled (FC) conditions. The onset of the transition into the superconducting state is found to occur at a critical temperature T_c of 3.4 K. The transition is quite sharp, which indicates good sample quality. The thermal irreversibility between the ZFC and the FC measurements provides evidence, that Pd_2HfAl is a type II superconductor. The large difference in magnitude between the ZFC and the FC measurements, which indicates a comparatively weak Meissner effect, results from flux pinning that is caused by bulk inhomogeneities.

The lower panel shows a plot of the field-dependent magnetization (butterfly loop) of Pd_2HfAl. The induction field was varied from -100 mT to 100 mT at a temperature of 2 K. The critical magnetic field H_{c1} at this temperature cannot be accurately determined because of a broadening of the magnetization maximum. In addition, flux pinning leads to a more symmetric shape of the curve, which impedes accurate determination of H_{c1}. The magnetization measurements for the corresponding Pd_2ZrAl are comparable to the magnetization curves shown for Pd_2HfAl and indicate that it is also a type II superconductor. For Pd_2ZrAl the onset of the superconducting transition was observed at a temperature of 3.2 K.

Figure 6.3 displays the results of the resistance measurements for both the Pd_2HfAl and Pd_2ZrAl superconductors close to the critical temperatures. Distinct drops of the resistance point on the transitions to the superconducting states. For both compounds, the resisitive superconducting transitions appear at slightly higher temperatures than the transitions determined from the magnetic measurements. This fact is well known and explained by bulk inhomogeneities.

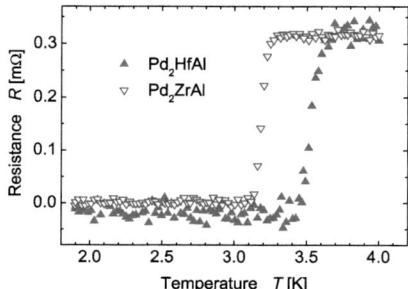

Figure 6.3: Low-temperature resistance measurements for Pd_2HfAl and Pd_2ZrAl.

Conclusion

In conclusion, starting with electronic structure calculations the Heusler compounds Pd_2HfAl and Pd_2ZrAl have been found as candidates for superconductivity in accordance with the van Hove scenario. These proposed superconducting candidates were synthesized by arcmelting and experimentally investigated. Superconductivity was found in both compounds. The critical temperatures T_c were determined to be 3.4 K and 3.2 K, respectively. Because of their low transition temperatures however, neither compound is at present very well suited for design of magnetization-controlled Josephson junctions but may be used for demonstration. It appears, however, that it may be possible to design a spin injector based purely on Heusler compounds using a superconductor along with a suitable half-metallic ferromagnet. Very high magnetic transition temperatures have already been realized in half-metallic Heusler compounds [77]. The task now is to find Heusler superconductors with higher transition temperatures so that magnetization-controlled Josephson junctions for applications, which are exclusively based on Heusler compounds, can be realized.

Acknowledgements

The authors would like to thank Martin Jourdan and Gerhard Jakob for assistance with the measurements, for many suggestions and for fruitful discussions.

7 Superconductivity in palladium-based Heusler compounds

The text of this chapter is identical with the following publication:
J. Winterlik, G. H. Fecher, A. Thomas, and C. Felser
Phys. Rev. B **79**, 064508 (2009).

Abstract

This work reports on four more Heusler superconductors: Pd_2ZrAl, Pd_2HfAl, Pd_2ZrIn, and Pd_2HfIn. These compounds exhibit superconducting transition temperatures ranging from $2.4 - 3.8$ K as determined by resistivity measurements. According to their behavior in an external magnetic field, all compounds are type II bulk superconductors. The occurrence of superconductivity was predicted for these compounds using electronic structure calculations. The electronic structures exhibit van Hove singularities (saddle points) at the L point. These lead to a maximum in the corresponding density of states and superconductivity according to the van Hove scenario. The superconducting properties of electron-doped and hole-doped substituted compounds $Pd_2B_{1-x}B'_xAl$, whereby $B=$Zr and Hf and $B'=$Y, Nb, and Mo, were investigated to obtain information about the dependence of the transition temperature on the density of states at the Fermi energy following the van Hove scenario. The calculated electronic structure reveals that the substituted compounds do not follow a rigid-band model. In addition, the random distribution of the substituted atoms strongly increases impurity-type electron scattering. The substituent concentrations used in this work lead to strongly enhanced impurity-type scattering and eventually to suppression of the superconducting state.

Introduction

Heusler compounds with the $L2_1$ structure type and a stoichiometric composition A_2BC became popular with the archetype Cu_2MnAl [78], a remarkable compound that shows ferromagnetic ordering despite the absence of any ferromagnetic element. At present, the Heusler family is widely associated with the research area of spintronic applications. Some Heusler compounds have been reported to show half-metallic ferromagnetism [62, 63], a feature of utmost concern with respect to the desired high degree of spin polarization, and thus for implementation in spintronic applications. Several Co-based Heusler compounds were employed as electrodes in magnetic tunnel junctions [64, 65]. In contrast to the magnetic Heusler compounds, Heusler superconductors play a minor role with respect to application because of their low transition temperatures. Investigating what causes a certain Heusler compound to show magnetic order or superconductivity is, however, of general scientific interest.

Up to now, very few Heusler compounds have been reported to be superconductors; the first of these were published in Ishikawa's pioneering review [66]. Currently, Pd_2YSn is the Heusler compound with the highest critical temperature T_c of 4.9 K [67]. Despite the occurrence of ferromagnetic ordering in elemental nickel, Heusler compounds with nickel at the Wyckoff $8c$ position were found to exhibit superconductivity; the compound with the currently highest critical temperature of 3.4 K, Ni_2NbSn, was reported in 1983 [67]. The coexistence of superconductivity and antiferromagnetic order was found in Heusler compounds Pd_2YbSn (Ref. [68]) and Pd_2ErSn [69].

For a better understanding of the magnetic properties of solid materials, we need to understand their electronic structures. We have performed electronic structure calculations using *ab initio* methods in order to find promising palladium-based Heusler compounds as candidates for superconductivity that match our design criterion, i.e., a saddle point at a certain high-symmetry point in the energy dispersion curve at the Fermi energy ϵ_F. Such a saddle point leads to a high density of states $n(\epsilon_F)$ at the Fermi energy. These are referred to as van Hove singularities [14]. Prospective candidates for superconductivity include certain Heusler compounds with 27 electrons that exhibit a saddle point at the L point close to ϵ_F in the corresponding energy dispersion curve according to the van Hove scenario [11, 12, 15]. It is known from the Bardeen-Cooper-Schrieffer (BCS) theory for superconductivity that the transition temperature of a superconducting material increases exponentially with an increasing $n(\epsilon_F)$ given that Debye frequency and Cooper-pairing interaction are independent of $n(\epsilon_F)$.[70] The unusually high transition temperatures of the intermetallic $A15$ superconductors, which are highly important for application purposes (i.e., as superconducting magnets), were explained with the van Hove scenario. Based on the van Hove scenario, we have already reported Heusler superconductors with 27 electrons: Pd_2ZrAl,

7. Superconductivity in palladium-based Heusler compounds

Pd_2HfAl, and Ni_2ZrGa [11, 12]. Here we present a more detailed investigation of the palladium-based compounds and report two more 27-electron superconducting Heusler compounds: Pd_2ZrIn and Pd_2HfIn. Electron-doping and hole-doping experiments were performed with Pd_2ZrAl and Pd_2HfAl to investigate the dependence of T_c on the energy of the corresponding van Hove singularities. To do this, we prepared and characterized solid solutions of pure compounds of $Pd_2B_{1-x}B'_xAl$, whereby B=Zr and Hf and B'=Y, Nb, and Mo.

Calculational details

The electronic and vibrational properties of the pure compounds were calculated by means of WIEN2k (Refs. [79] and [74]) in combination with PHONON [80]. The electronic properties of the substituted - electron-doped or hole-doped - compounds were calculated using the Korringa-Kohn-Rostocker method together with the coherent potential approximation [81, 82].

The electronic structure of the pure A_2BC compounds was calculated using the full potential linearized augmented plane-wave (FLAPW) method as implemented in WIEN2k. The exchange-correlation functional was taken within the generalized gradient approximation (GGA) [75]. A $25 \times 25 \times 25$ point mesh was used as basis for integration in the cubic systems resulting in 455 k points in the irreducible wedge of the Brillouin zone. The energy convergence criterion was set to 10^{-5} Ry while, at the same time, the criterion for charge convergence was set to $1 \times 10^{-3} e^-$. The muffin-tin radii were set to 2.5 a_{0B} ($a_{0B} = 0.5291772$ Å) for the transition metals as well as the main group element. The primitive fcc cell, which contains four atoms, was enlarged to a cell with 16 distinguished atoms to calculate the Hellmann-Feynman forces for the phonon analysis. For these calculations, a force convergence criterion of 1×10^{-4} Rya_{0B}^{-1} was used in addition to the energy and charge convergence criteria. All phonon calculations were performed for the optimized lattice parameter.

The electronic structure of the substituted $A_2B_{1-x}B'_xC$ compounds was calculated by means of the full relativistic Korringa-Kohn-Rostocker (SPRKKR) method [81, 82]. The exchange-correlation functional was taken within the Vosko-Wilk-Nusair parameterization [83, 84]. No noticeable differences in the electronic structure were observed when using the full potential version of the program together with GGA. The coherent potential approximation (CPA) was used to account for the random distribution of the B and B' atoms in the $Pd_2B_{1-x}B'_xAl$ or $Pd_2B_{1-x}B'_xIn$ solid solutions.

The electronic structures reported below (see Sec. 7) were calculated for the experimental as well as the optimized lattice parameter of each compound. Figure 7.1 shows the results for the optimization of the lattice parameter for Pd_2ZrAl and Pd_2HfAl. The optimized lattice parameters from an equation of states fit [76] are slightly larger

Table 7.1: Lattice parameters of palladium-based superconducting Heusler compounds. Compared are the experimental values a_{exp} (see Sec. 7) with the optimized values a_{opt}. B is the bulk modulus.

	a_{exp} [Å]	a_{opt} [Å]	B [GPa]
Pd_2ZrAl	6.406	6.467	151
Pd_2ZrIn	6.536	6.623	141
Pd_2HfAl	6.376	6.436	159
Pd_2HfIn	6.530	6.592	150

than experimental parameters by about 1%. This is common for calculations in the generalized gradient approximation. GGA slightly overestimates the lattice parameter compared to the pure local-density approximation, which tends to overbinding [85–87]. The accompanying bulk moduli for Pd_2ZrAl and Pd_2HfAl are 151 and 159 GPa, respectively. The results from the optimization of the lattice parameters are summarized in Table 7.1.

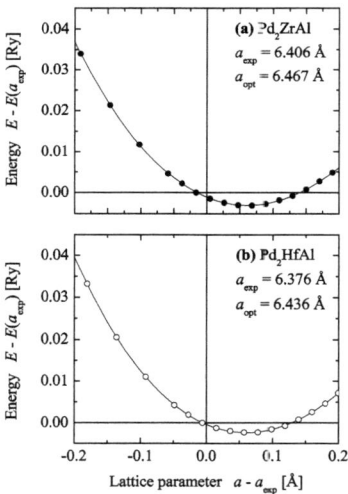

Figure 7.1: Optimization of the lattice parameter. Shown is the optimization for Pd_2ZrAl in (a) and Pd_2HfAl in (b). The solid lines are the results of an equation of state fit.

Experimental details

Polycrystalline ingots were prepared by repeated arc melting of stoichiometric mixtures of the elements in an argon atmosphere at a pressure of 1×10^{-4} mbar to avoid oxygen contamination. The samples were subsequently annealed for two weeks at 1073 K in evacuated quartz tubes. After the annealing process, the samples were quenched to 273 K in a mixture of ice and water to retain the desired $L2_1$ structure. Slow cooling of the samples after the annealing process leads to enhanced $B2$-type anti-site disorder. This is mainly a disorder between the transition metal (B) on the $4a$ site with the main group element (C) on the $4b$ site that is often found in Heusler compounds.

The crystal structures of the compounds were investigated using powder x-ray diffraction (XRD). The XRD measurements were carried out using a Siemens D5000 diffractometer with monochromatized Cu K_α radiation. The purity of the compounds was confirmed using scanning electron microscopy with energy dispersive x-ray spectroscopy (EDX). The superconducting transitions were verified in resistivity measurements using a Physical Property Measurement System (PPMS, Quantum Design, Model 6000). The resistivity of samples with polished surfaces was measured using the four-point probe technique. The diamagnetic shielding and the Meissner effect of the compounds were investigated using temperature-dependent and field-dependent magnetization measurements with a superconducting quantum interference device (SQUID, Quantum Design, MPMS-XL-5). Small spherical sample pieces of approximately 20 to 120 mg were used to ensure precise measurements. The temperature-dependent magnetization measurements were carried out under zero-field-cooled (ZFC) and field-cooled (FC) conditions. For the ZFC measurements, the samples were first cooled down to 1.8 K without applying any magnetic field. Afterward, an external magnetic field of 2.5 mT was applied, and the sample magnetization was measured with increasing temperature. The FC magnetization was recorded subsequently in the same induction field upon cooling down to 1.8 K again. The field-dependent magnetization was recorded at a temperature of 2 K, well below the transition temperatures of all superconductors.

Results and discussion

Pure compounds

Structural characterization

All compounds crystallize in the cubic $L2_1$ Heusler structure (space group: $Fm\bar{3}m$), where the Wyckoff positions are $8c$ $(\frac{1}{4},\frac{1}{4},\frac{1}{4})$ for Pd atoms, $4a$ (0,0,0) for B=Zr or Hf atoms and $4b$ $(\frac{1}{2},\frac{1}{2},\frac{1}{2})$ for C=Al or In atoms. Figure 7.2 shows the diffraction patterns for all pure compounds. The raw data (black) are compared to the difference between a calculated Rietveld refinement and the raw data (gray). The experimental lattice parameters as found from Rietveld refinements are given in Table 7.1. The patterns indicate excellent phase purity for all compounds. In the case of Pd_2ZrIn, the (111) reflex is very small, and the (200) reflex is not seen at all. This is not an effect of $B2$-type disorder but is instead due to almost equal scattering factors of Pd, Zr, and In. Even in a presumably perfectly ordered $L2_1$ structure of Pd_2ZrIn, the intensity of the (200) reflex would be less than 1% of that of the (220) reflex.

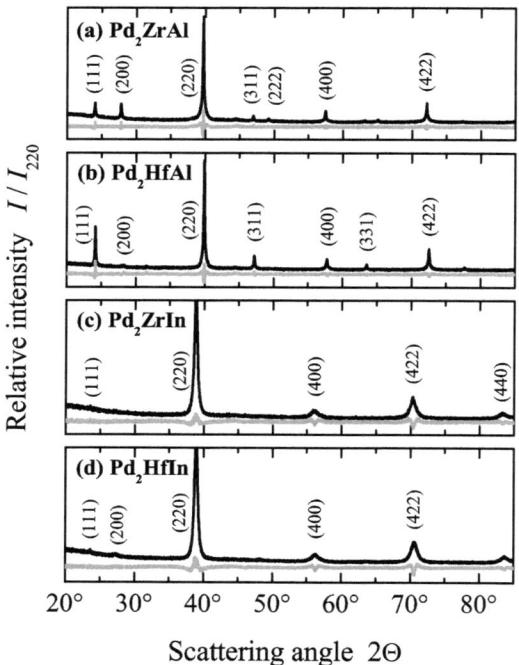

Figure 7.2: X-ray diffraction patterns of the pure Heusler superconductors. (a)-(d) display the patterns for Pd_2ZrAl, Pd_2HfAl, Pd_2ZrIn, and Pd_2HfIn.

Hf and Zr are very difficult to separate during purification of the elements even in high-purity materials. The used Zr is stated to contain less than 3% Hf by the manufacturer and vice versa, < 3% Zr in Hf. To detect this type of cross contamination, polished disks of the materials were investigated using scanning electron microscopy with EDX. No traces of Hf were found in the Zr-containing compounds and vice versa as shown in Fig. 7.3. According to the methodological detection limit, the impurities of Pd_2HfIn in Pd_2ZrIn are $\leq 1\%$.

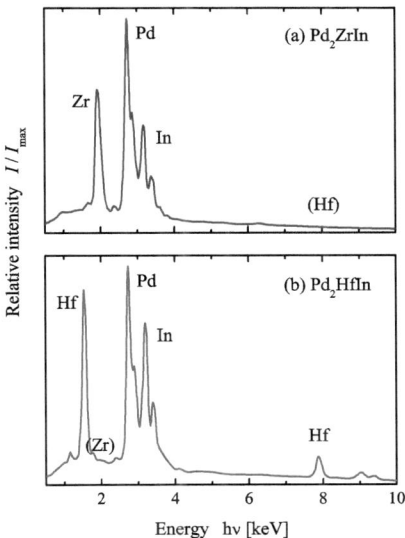

Figure 7.3: EDX spectra of (a) Pd_2ZrIn and (b) Pd_2HfIn. No cross contamination of Zr and Hf is observed.

Calculational results

The calculated electronic structures of Pd_2ZrAl and Pd_2ZrIn are shown in Figs. 7.4 and 7.5. The electronic structure of both compounds is very similar. The hybridization gap - typical for Heusler compounds - that separates the low-lying s states from the d bands is clearly visible at around 6 eV below the Fermi energy in the energy dispersion curve as well as in the density of states of both compounds. This gap is slightly larger in Pd_2ZrIn, reflecting the larger lattice parameter as compared to Pd_2ZrAl.

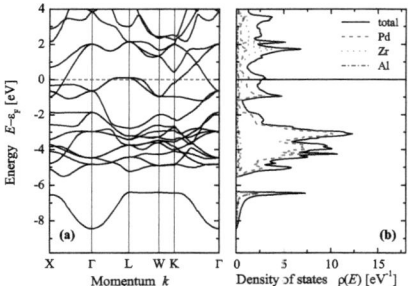

Figure 7.4: Electronic structure of Pd_2ZrAl. (a) displays the band structure and (b) the density of states. (Both were calculated using FLAPW for the optimized lattice parameter.)

Most interesting, however, is the occurrence of a van Hove singularity just above the Fermi energy at the L point. At the same time, only one rather steep, d band is crossing the Δ direction. The flat bands forming the van Hove singularity at the L point result in a maximum density of states. This maximum density emerging from the states just above ϵ_F is about 3.5 eV^{-1} in Pd_2ZrAl and thus about 0.8 eV^{-1} higher than in Pd_2ZrIn. In both compounds, the density is slightly lower directly at ϵ_F.

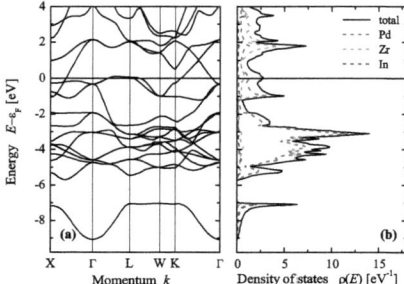

Figure 7.5: Electronic structure of Pd_2ZrIn. (a) displays the band structure and (b) the density of states. (Both were calculated using FLAPW for the optimized lattice parameter.)

The energy of the states at the high-symmetry points is summarized in Table 7.2 together with the density of states $n(\epsilon_F)$ at the Fermi energy. Table 7.2 shows that the

Table 7.2: Energy of the van Hove singularities close to ϵ_F. The experimental lattice parameters a of the Al-containing compounds are assigned by *. All energies E are given with respect to the Fermi energy ϵ_F.

	a [Å]	$E(L)$ [meV]	$E(\Gamma)$ [meV]	$E(X)$ [meV]	$n(\epsilon_F)$ [eV^{-1}]
Pd$_2$ZrAl	6.406*	108	- 694	- 317	2.6
	6.467	109	- 630	- 299	2.7
Pd$_2$ZrIn	6.623	65	- 307	- 270	2.7
Pd$_2$HfAl	6.376*	172	- 643	- 287	2.3
	6.436	173	- 585	- 272	2.4
Pd$_2$HfIn	6.592	136	- 171	- 245	2.4

energy of the van Hove singularity at the L point is rather resistant to variation in the lattice parameter, whereas some deviations are found in the energies of the states at the Γ and X points if comparing the values calculated for the experimental and optimized lattice parameters. The density of states at the Fermi energy is slightly higher in the Zr compounds compared to the Hf compounds. Overall, no remarkable differences are noticed if comparing the compounds or the different lattice parameters used in the calculations.

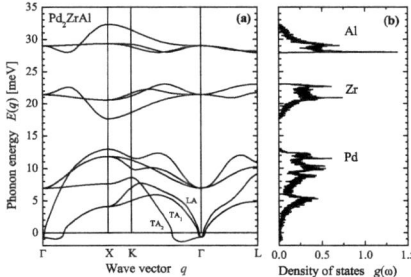

Figure 7.6: Vibrational structure of Pd$_2$ZrAl. (a) displays the phonon dispersion and (b) the phonon density of states $g(\omega)$. Note that negative energies correspond to solutions of the wave equation with complex q values. (The calculations were performed with the optimized lattice parameter.)

Figure 7.6 shows the phonon dispersion and the phonon spectrum of Pd$_2$ZrAl. The vibronic structure exhibits four groups of vibrational states, the low-lying acoustical modes, and the high-lying optical modes. For high-symmetry directions, the set of acoustic modes (LA, TA_1, and TA_2) is non-degenerate only along Γ-K. The optical modes are three-fold degenerate at Γ as is revealed in the splitting of the bands in the

Σ direction. The highest optical modes originate from the Al atoms and the lowest optical modes arise from Pd (see Fig. 7.6(b)).

The acoustical modes of Pd_2ZrAl are unstable close to the Γ point. The energies with negative values correspond to wave vectors with complex q values. This instability is most pronounced in the TA_2 mode, which is unstable over a wider range of q. Such instabilities were previously observed in Ni-based and Co-based Heusler compounds [88, 89]. For the example of Ni_2TiGa, it was shown [89] that magnetic order is not a necessary condition for phonon softening to occur. Here, this fact is demonstrated for another nonmagnetic compound.

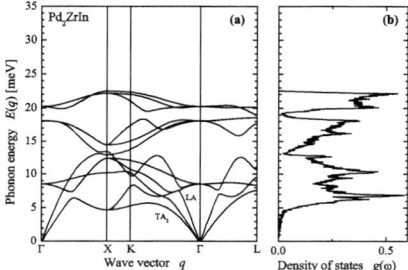

Figure 7.7: Vibrational structure of Pd_2ZrIn. (a) displays the phonon dispersion and (b) the phonon density of states. (The calculations were performed with the optimized lattice parameter.)

The phonon dispersion and density of Pd_2ZrIn is shown in Fig. 7.7, in which the density can be seen to cover a smaller range of phonon energies. Due to the higher mass of In, the splitting of the three groups of optical modes is less pronounced compared to Pd_2ZrAl. The gaps between the groups are closed or become pseudogaps.

Table 7.3 summarizes the energies of the optical phonon modes and compares the average phonon frequency $\overline{\omega} = \int \omega g(\omega) d\omega$ to the Debye temperature Θ_D. Θ_D was determined from a fit of the calculated specific heat of the phonons to a Debye model. The high-lying optical modes at about 30 meV in the Al-containing compounds do not overlap with the remainder of the phonon spectrum and appear like Einstein frequencies. In a hybrid Einstein-Debye model for Pd_2HfAl, the corresponding Einstein temperature is $\Theta_E \approx 350$ K calculated from the average phonon frequencies of that mode. The Debye temperature of Pd_2ZrAl was not clearly revealed from the fit. This is partially due to the observed phonon softening. Most probably, the clear splitting of the Zr-induced optical mode requires a hybrid model with two different Einstein

7. Superconductivity in palladium-based Heusler compounds

Table 7.3: Energetics of the optical phonon modes and average temperatures. The energy at X corresponds to the highest mode. The temperature equivalent of the mean phonon frequency $\overline{\omega}$ is given for better comparison to the Debye temperature Θ_D [note that the values with * are not accurate due to the phonon softening in Pd_2ZrAl (see Fig. 7.6)].

	$\hbar\omega(\Gamma)_1$ [meV]	$\hbar\omega(\Gamma)_2$ [meV]	$\hbar\omega(\Gamma)_3$ [meV]	$\hbar\omega(X)$ [meV]	$\overline{\omega}$ [K]	Θ_D [K]
Pd_2ZrAl	6.96	21.4	29.0	32.3	194*	> 310*
Pd_2ZrIn	8.52	18.1	20.2	22.5	159	220
Pd_2HfAl	10.2	18.9	28.5	31.1	194	277
Pd_2HfIn	9.73	16.5	18.8	21	153	209

temperatures to describe the specific heat correctly. The Debye temperatures of the In-containing compounds are lower compared to those with Al, as expected from the differences of the phonon spectra.

Resistivity

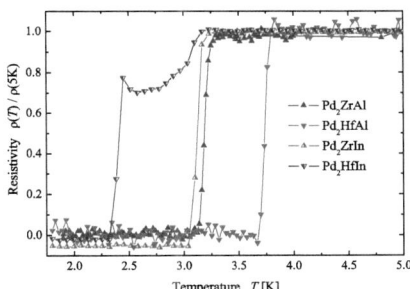

Figure 7.8: Resistive superconducting transitions of the pure compounds. Shown is the temperature-dependent resistivity in the region around T_c^{mid} for the Heusler superconductors Pd_2ZrAl, Pd_2HfAl, Pd_2ZrIn, and Pd_2HfIn.

The superconducting transitions of the pure compounds were verified with temperature-dependent resistivity measurements. Figure 7.8 shows the resistive superconducting transitions of the pure compounds. The transition temperatures T_c^{mid} were determined by calculating the mean value of the onsets and offsets of the superconducting transitions. The values are summarized in Table 7.4. In the case of Pd_2HfIn, the onset of a superconducting transition can already be observed at approximately 3.1 K, but the resistivity does not drop straight to zero. This jump is then followed by the pronounced bulk superconducting transition for the Heusler compound at 2.4 K. No impurities were found in the XRD pattern of Pd$_2$HfIn. It is therefore concluded that this broad onset may be attributed to bulk inhomogeneities of the sample. It may be caused by an unidentified ternary or binary impurity phase below the XRD detection limit consisting of the elements Pd, Hf, and In as neither contaminations nor impurities of Zr were found in the EDX analysis.

Magnetic properties

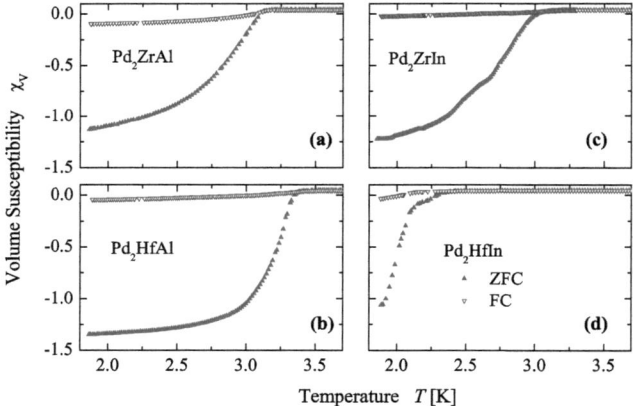

Figure 7.9: Temperature-dependent magnetization of the pure compounds. (a)-(d) show the volume susceptibilities χ_V in the regions around T$_c$ upon ZFC and FC conditions for Pd$_2$ZrAl, Pd$_2$HfAl [11], Pd$_2$ZrIn, and Pd$_2$HfIn. The measurements were performed with magnetic induction fields $\mu_0 H$ of 2.5 mT, respectively.

The results of the temperature-dependent magnetization measurements are given in Figs. 7.9(a)-(d). All compounds exhibit superconducting transitions. The onsets of these transitions are given in Table 7.4. Thermal irreversibilities between ZFC and FC measurements indicate that all compounds are type II superconductors. All samples show quite sharp superconducting transitions indicating low grades of impurity. However, very large differences in magnitude between ZFC and FC measurements are found in all cases indicating a reduced Meissner effect. This feature is caused by flux pinning that is attributed to bulk inhomogeneities. For all compounds, the resistive superconducting transitions appear at slightly higher temperatures as compared to the magnetization measurements. This fact is well known. The resistive transition takes place when one percolation path throughout the sample becomes superconducting. The magnetic transition requires a certain superconducting volume. The volume susceptibilities χ_V were calculated assuming a demagnetization factor of $\frac{1}{3}$ of a perfect sphere. Using this approximation for all compounds confirms bulk superconductivity. Precise statements about the superconducting volumes, however, cannot be made because the measured samples do not correspond to perfect spherical or other geometrical bodies. The segregated unidentified impurity, which was found in the resistivity measurement of Pd_2HfIn, is also observed as additional structure in the corresponding temperature-dependent magnetization curve. The magnetic data of Pd_2ZrIn also exhibit small deviations. This suggests that the unidentified impurity may consist of the elements Pd and In only. Due to the higher transition temperature of Pd_2ZrIn, this effect is not visible in the corresponding resistivity measurement. The critical temperatures of both compounds may increase if formation of this impurity phase can be avoided.

Table 7.4: Critical temperatures of the palladium-based Heusler superconductors. The critical temperatures T_c^{mid} of the resistivity measurements represent mean values of the superconducting onsets and offsets. The critical temperatures from the magnetization measurements T_c^{Mag} represent the onsets of the superconducting transitions.

	T_c^{Res} [K]	T_c^{Mag} [K]	H_{c1} [mT]
Pd_2ZrAl	3.2	3.1	6.0
Pd_2ZrIn	3.1	3.0	12.3
Pd_2HfAl	3.8	3.4	10.3
Pd_2HfIn	2.4	2.3	6.2

Figures 7.10(a)-(d) shows the field-dependent magnetizations (butterfly curves) at temperatures of 2 K in the superconducting state for all compounds. Theoretically, the critical magnetic field H_{c1} for a certain temperature and a certain compound can be determined from the virgin curve of the field-dependent magnetization. This is not possible, however, for these intermetallic samples, because the bulk inhomogeneities lead to a more symmetric shape of the curves, which impedes accurate calculations of H_{c1}. Rough estimations provide values of H_{c1} for the compounds as presented in Table 7.4.

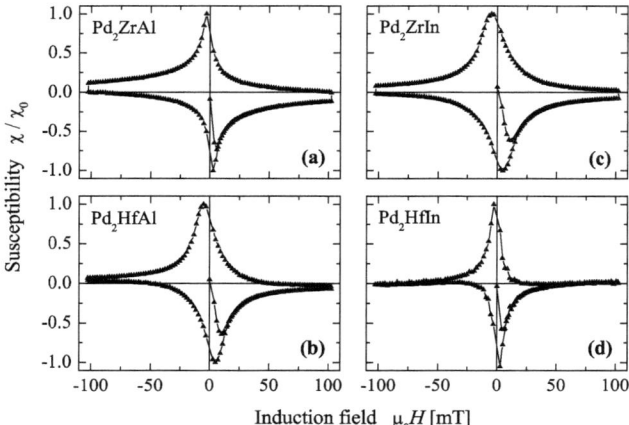

Figure 7.10: Butterfly curves for Pd_2ZrAl, Pd_2HfAl [11], Pd_2ZrIn, and Pd_2HfIn. (a)-(d) show the normalized susceptibilities χ/χ_0 at temperatures of 2 K in a range from -100 to 100 mT, respectively. All curves were measured from the ZFC state.

Substituted compounds

According to the van Hove scenario [15], a maximum T_c should be obtained for a maximum $n(\epsilon_F)$ at ϵ_F at a given Debye frequency and Cooper-pairing interaction. For the cases of Pd_2ZrAl and Pd_2HfAl, the van Hove singularities coincide with ϵ_F when the compounds are doped with electrons. Assuming a rigid-band model and a fixed lattice parameter, the calculated electronic structure suggests that doping by about 0.1 of an electron shifts the band with the singularity at L onto ϵ_F. This may be established by substituting the group 4 element (Zr, Hf) by 10% of an element of group 5 (V, Nb) or 5% in case of group 6 elements (Cr, Mo). In a similar way, a substitution by a group 3 (Sc, Y) element will shift ϵ_F downwards to the singularity at X. To investigate the behavior of T_c upon electron doping and hole doping the solid solutions of $Pd_2B_{1-x}B'_xAl$ with B=Zr and Hf and B'=Y, Nb, and Mo were prepared as described in Sec. 7. For Pd_2ZrAl, Y and Nb could be substituted for Zr, whereas only Y could be substituted for Hf in Pd_2HfAl. Doping with Mo was not possible for either compound.

Structural characterization

The crystal structures of the substituted alloys were analyzed as described in Sec. 7. The alloys crystallize in the cubic $L2_1$ Heusler structure (space group: $Fm\bar{3}m$). With the exception of $Pd_2Hf_{0.5}Y_{0.5}Al$, none of the alloys exhibit any impurities in the XRD patterns. The experimental lattice parameters are obtained from Rietveld refinements (see Table 7.5). The increasing and decreasing lattice parameters in the solid solutions originate from the differing atomic radii of the substituted elements. The atomic radii of Nb and Mo are smaller than those of Zr and Hf, whereas the atomic radius of Y is larger. This is reflected in the lattice parameters listed in Table 7.5. Besides the impurities, which were found in the XRD pattern of $Pd_2Hf_{0.5}Y_{0.5}Al$, the small difference between the lattice parameters of $Pd_2Hf_{0.75}Y_{0.25}Al$ and $Pd_2Hf_{0.5}Y_{0.5}Al$ supports the conclusion that the lattice of $Pd_2Hf_{1-x}Y_xAl$ becomes saturated with Y when $0.25 \leq x \leq 0.5$. Increasing the Y concentration above this limit causes impurities to segregate. One of these impurities was identified as elemental yttrium.

Calculational results

Figure 7.11 shows a comparison between the density of states of the $Pd_2Zr_{0.9}Y_{0.1}Al$ and $Pd_2Zr_{0.9}Nb_{0.1}Al$ solid solutions and of Pd_2ZrAl in its pure form. The calculations were performed using SPRKKR and the results agree very well with the density of states from the FLAPW calculations. The solid solutions were treated in the CPA approximation. It should be noted that a band structure is no longer persistent in the random alloys due to the lack of periodicity.

Figure 7.11(a) shows that the maximum density of states close above ϵ_F, being indicative for the van Hove singularity at the L point, is slightly shifted to higher energies in the Y-substituted compound. This is due to underdoping of electrons. The position of the maximum is rather stable for electron doping in the Nb-containing compound (see Fig. 7.11(c)) in comparison to the pure Zr compound (see Fig. 7.11(b)). This shows that bands do not behave completely rigid. However, a small shift of ϵ_F is evident. Adequate electron doping should thus allow the van Hove singularity to coincide with ϵ_F. In theory, the situation is rather similar for electron-doped and hole-doped Pd_2HfAl.

Figure 7.12 illustrates the effect of electron doping on the electronic properties for the examples of $Pd_2Zr_{1-x}Nb_xAl$ and $Pd_2Hf_{1-x}Nb_xAl$. The energy shift of the maximum of $n(E)$ associated with the van Hove singularity at the L point cannot be given more precisely due to lack of a band structure. From the fit, it is expected that the singularity coincides with ϵ_F at a Nb concentration of approximately 20% or 30% in the alloys based on Zr or Hf, respectively. The density of states $n(\epsilon_F)$ at the Fermi energy increases with increasing Nb content.

A detailed analysis of Pd_2HfAl reveals that this increase is not primary due to the shift of the maximum of the density of states arising from Pd states of the pure compound but by an increase in the density located at the Nb atoms. This reordering of the electron densities demonstrates that a rigid-band model is not directly applicable for this sort of compounds even if using a fixed lattice parameter in the calculations. Pd_2ZrAl shows an intermediate maximum of $n(\epsilon_F)$. In this case, the local density of states $n_{Pd}(\epsilon_F)$ at the Pd atoms also exhibits a maximum. Here, the Nb concentration x at the intermediate maximum of $n(x, \epsilon_F)$ corresponds to the one where the van Hove saddle point at L coincides with the Fermi energy.

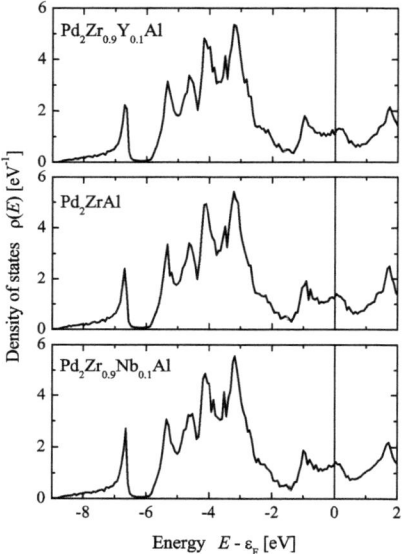

Figure 7.11: Electronic structure of $Pd_2Zr_{0.9}B_{0.1}Al$, B=Y, Nb. Compared is the density of states of the pure compound to the cases of 10% substitution of Zr by Y or Nb. (Calculated by SPRKKR with the experimental lattice parameter.)

In a trivial analysis, the increase in states at ϵ_F may suggest that the resistivity decreases already in the normal conducting state. This is, indeed, not the case due to the strong influence of impurity scattering in the substituted compounds. The random distribution of the Zr or Hf and Nb atoms on a common site of the lattice leads to an increase in the resistivity as is also observed for simple binary alloys such as

7. Superconductivity in palladium-based Heusler compounds

$Pd_{1-x}Ag_x$ [90, 91]. It is conceivable that reordering of charges and/or impurity scattering prevent an increase in the superconducting transition temperature.

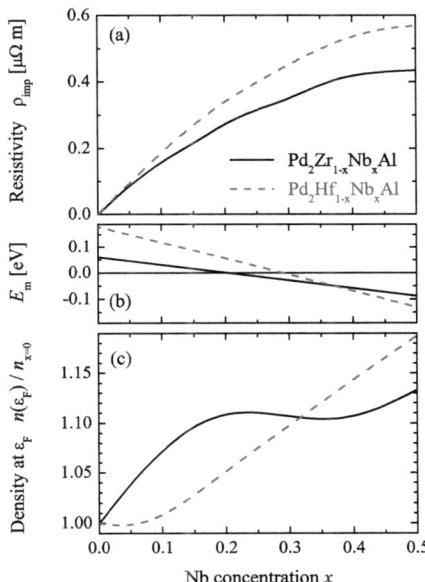

Figure 7.12: Electron doping in $Pd_2Zr_{1-x}Nb_xAl$ and $Pd_2Hf_{1-x}Nb_xAl$. (a) shows the residual resistivity due to *impurity* scattering, (b) shows the energy shift of the maximum of $n(E)$ associated with the van Hove singularity at the L point, and (c) shows the density of states at the Fermi energy relative to the density in the pure compounds with $x = 0$. (Calculated for the *normal* conducting state by SPRKKR with the fixed lattice parameters of the pure compounds.)

Resistivity

All solid solutions were investigated with respect to superconductivity in temperature-dependent resistivity measurements as described in Section 7. The corresponding critical temperatures T_c^{mid} are given in Table 7.5. Figure 7.13 shows the resistivity measurement of Pd_2ZrAl in comparison to the alloys $Pd_2Zr_{0.9}Y_{0.1}Al$ and $Pd_2Zr_{0.9}Nb_{0.1}Al$. A superconducting transition was only found in $Pd_2Zr_{0.9}Nb_{0.1}Al$ ($T_c^{mid} = 2.1$ K). In this alloy, resistivity already begins to decrease at approximately 4.1 K. This may be attributed to a superconducting impurity phase below the detection limit of XRD. This phase may be a disordered cubic $A15$ material [92]. Some of these are known to exhibit superconducting transitions around 4 to 5 K. $Pd_2Zr_{0.9}Y_{0.1}Al$ does not show a superconducting transition. A small drop in resistivity occurs at approximately 2.6 K, but this does not correspond to a bulk superconducting transition. Most probably, this drop is caused by small superconducting paths in regions with lower yttrium contents throughout the inhomogeneous bulk sample. Figure 7.14 shows the temperature-dependent resistivity of Pd_2HfAl in comparison to the alloys $Pd_2Hf_{0.75}Y_{0.25}Al$ and $Pd_2Hf_{0.5}Y_{0.5}Al$. Superconducting transitions are observed in both alloys but the transition temperature decreases with increasing yttrium concentration.

Table 7.5: Properties of the substituted palladium-based Heusler superconductors. a are the measured lattice parameters and T_c^{mid} is the critical temperatures from resistivity measurements.

	a [Å]	T_c^{mid} [K]
$Pd_2Zr_{0.9}Y_{0.1}Al$	6.432	-
$Pd_2Zr_{0.9}Nb_{0.1}Al$	6.372	2.1
$Pd_2Hf_{0.75}Y_{0.25}Al$	6.393	3.2
$Pd_2Hf_{0.5}Y_{0.5}Al$	6.404	2.9

Maximum transition temperatures were expected for electron-doped Pd_2ZrAl and Pd_2HfAl according to the van Hove scenario. For Pd_2HfAl, no conclusion can be drawn about the possibility of electron doping and increasing T_c because only Y was incorporated successfully into the unit cell of Pd_2HfAl. This hole doping led to a decrease in T_c with increasing Y content as expected from the electronic structure. For Pd_2ZrAl, hole doping with 10% of Y led to destruction of superconductivity, and the superconducting transition is lowered in the alloy $Pd_2Zr_{0.9}Nb_{0.1}Al$ as compared to Pd_2ZrAl. The expected behavior was thus not found. Nonmagnetic scattering should not suppress superconductivity by itself according to Anderson's theorem [93]. It was, however, shown that enhanced disorder leads to spatial fluctuations of the superconducting gap Δ and eventually to suppression of the superconducting state [94]. Thus, a Nb concentration of 10% manifestly provokes a degree of disorder, which is sufficient to reduce T_c considerably. Electron-doping experiments with the cubic $A15$ superconductors led to similar results, where T_c is very often lowered upon electron doping or hole doping via element substitution [72].

7. Superconductivity in palladium-based Heusler compounds

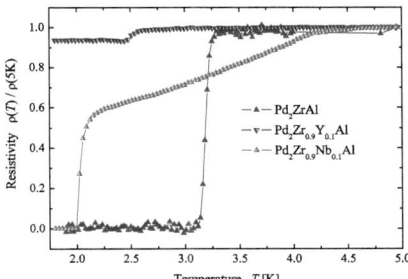

Figure 7.13: Resistive superconducting transitions for the solid solutions $Pd_2Zr_{1-x}B'_xAl$. Shown is the temperature-dependent resistivity in the region around T_c^{mid} for the pure compound Pd_2ZrAl as well as the alloys $Pd_2Zr_{0.9}Y_{0.1}Al$ and $Pd_2Zr_{0.9}Nb_{0.1}Al$.

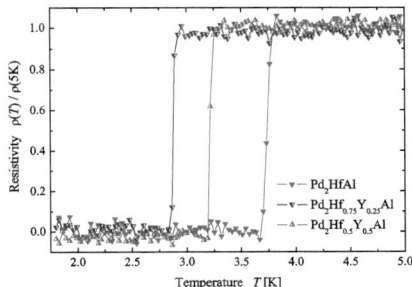

Figure 7.14: Resistive superconducting transitions for the solid solutions $Pd_2Hf_{1-x}Y_xAl$. Shown is the temperature-dependent resistivity in the region around T_c^{mid} for the pure compound Pd_2HfAl as well as the alloys $Pd_2Hf_{0.75}Y_{0.25}Al$ and $Pd_2Hf_{0.5}Y_{0.5}Al$.

Summary and conclusions

Using *ab initio* calculations as a basis, the Heusler compounds Pd_2ZrAl, Pd_2HfAl, Pd_2ZrIn, and Pd_2HfIn have been identified as prospects for superconductivity according to the van Hove scenario. Resistivity and magnetization measurements provided evidence that all compounds are type II bulk superconductors with transition temperatures in the range of 2.4-3.8 K. Following the van Hove scenario, solid solutions of $Pd_2B_{1-x}B'_xAl$ with B=Zr and Hf and B'=Y, Nb, and Mo were synthesized to investigate the dependence of superconducting properties on the valence electron concentration and thus the density of states at the Fermi energy. Although the incorporation of some elements was possible for both compounds Pd_2ZrAl and Pd_2HfAl, superconductivity was suppressed in all solid solutions as compared to the pure compounds. The rather sizeable amounts of the substituents used here provoked a degree of disorder beyond the validity of the Anderson theorem. Instead of shifting the van Hove singularities to ϵ_F they caused suppressions of the superconducting state.

Acknowledgements

This work has been funded by the DFG (German Research Foundation) through the *"Condensed Matter Systems with Variable Many-Body Interactions"* Collaborative Research Center (Transregio SFB/TRR 49). The authors thank Changhai Wang and Andrei Gloskovskii for the EDX analysis and Gerhard Jakob for suggestions and fruitful discussions. The authors are also very grateful to P. Blaha (Vienna, Austria), H. Ebert (Munich), and their groups for development and providing the computer codes.

8 Ni-based superconductor: Heusler compound ZrNi$_2$Ga

The text of this chapter is identical with the following publication:
J. Winterlik, G. H. Fecher, C. Felser, M. Jourdan, K. Grube, F. Hardy, H. v. Löhneysen, K. L. Holman and R. J. Cava
Phys. Rev. B **78**, 184506 (2008).

Abstract

This work reports on the Heusler superconductor ZrNi$_2$Ga. Compared to other nickel-based superconductors with Heusler structure, ZrNi$_2$Ga exhibits a relatively high superconducting transition temperature of T_c = 2.9 K and an upper critical field of $\mu_0 H_{c2}$ = 1.5 T. Electronic structure calculations show that this relatively high T$_c$ is caused by a van Hove singularity, which leads to an enhanced density of states at the Fermi energy $N(\epsilon_F)$. The van Hove singularity originates from a higher-order valence instability at the L point in the electronic structure. The enhanced $N(\epsilon_F)$ was confirmed by specific-heat and susceptibility measurements. Although many Heusler compounds are ferromagnetic, our measurements of ZrNi$_2$Ga indicate a paramagnetic state above T_c and could not reveal any traces of magnetic order down to temperatures of at least 0.35 K. We investigated in detail the superconducting state with specific-heat, magnetization, and resistivity measurements. The resulting data show the typical behavior of a conventional, weakly coupled BCS (s-wave) superconductor.

8. Ni-based superconductor: Heusler compound ZrNi$_2$Ga

Introduction

In the research area of spintronic applications, Heusler compounds have become of interest as half-metals, where due to the exchange splitting of the d-electron states, only electrons of one spin direction have a finite density of states (DOS) at the Fermi level $N(\epsilon_F)$ [62, 63]. Up to the present, very few Heusler superconductors with the ideal formula of AB_2C have been found. In 1982, the first Heusler superconductors were reported, each with a rare-earth metal in the B position [66]. Among the Heusler superconductors, Pd-based compounds have attracted attention because YPd$_2$Sn exhibits the highest yet recorded T$_c$ of 4.9 K [67]. Moreover, coexistence of superconductivity and antiferromagnetic order was found in YbPd$_2$Sn (Ref. [68]) and ErPd$_2$Sn [69]. A systematic investigation of Ni-based Heusler compounds seems to be worthwhile as nickel has many properties in common with palladium but tends more toward magnetic order due to the smaller hybridization of the 3d states. In fact, elementary nickel is a ferromagnet. Thus, nickel-containing Heusler compounds with a high proportion of Ni are naively expected to show magnetic order rather than superconductivity. However, superconductivity of Ni-rich alloys NbNi$_2$C (C = Al, Ga, Sn) has been reported some time ago, with transition temperatures T_c ranging from 1.54 K to the highest recorded transition temperature of a Ni-based Heusler compound of 3.4 K in NbNi$_2$Sn [67, 95]. In contrast to the two aforementioned Pd-based compounds, these superconductors do not show indications of magnetic order. Currently there is a lot of excitement about the new high-temperature superconductors based on FeAs [96]. The superconductivity of these compounds is related to two-dimensional layers of edge-shared FeAs tetrahedrons [97]. These structure types can be understood as two-dimensional variants of the Heusler structure.

A clear understanding of the origin of superconductivity, magnetism, and their possible coexistence in Heusler compounds is still missing. To shed light on the relation between the electronic structure and the resulting ground state of AB$_2$C Heusler compounds we searched for another Ni-based Heusler compound with a high DOS at ϵ_F close to the Stoner criterion for ferromagnetism. A possible route for increasing $N(\epsilon_F)$ is the use of saddle points in the energy dispersion curves of the electronic structure. They lead to maxima in the DOS, which are the so-called van Hove singularities [14]. In order to identify such compounds, we have performed electronic structure calculations using *ab initio* methods. In a simple approach following the Bardeen-Cooper-Schrieffer (BCS) theory and neglecting any magnetic order, we would expect that the superconducting transition temperature of such compounds increases with $N(\epsilon_F)$ according to $T_c \approx \Theta_D \exp[-1/V_0 N(\epsilon_F)]$ if the Debye temperature Θ_D and the Cooper-pairing interaction V$_0$ are independent of $N(\epsilon_F)$. In fact, this van Hove scenario, where a maximum in the DOS is ideally located at ϵ_F, was used to explain the unusually high transition

8. Ni-based superconductor: Heusler compound ZrNi$_2$Ga

temperatures of the intermetallic A15 superconductors [98]. The correspondence between T_c and the valence electron count is known as the Matthias rule [71]. According to this rule, the high T_c of the A15 compounds was related to electron concentrations of about 4.6 and 6.4 electrons/atom, leading to a maximum of the DOS at ϵ_F [72]. On the basis of the van Hove scenario, we already found superconductivity in two Heusler compounds with 27 electrons: ZrPd$_2$Al and HfPd$_2$Al [11, 15]. Here, we report on the theoretical and experimental characterization of another Ni-containing superconducting Heusler compound ZrNi$_2$Ga. Additionally, electron-doped alloys Zr$_{1-x}$Nb$_x$Ni$_2$Ga were prepared and investigated to obtain information about the dependence of T_c on the location of the van Hove singularity.

Experimental details

Polycrystalline ingots of ZrNi$_2$Ga and electron-doped alloys Zr$_{1-x}$Nb$_x$Ni$_2$Ga were prepared by repeated arc melting of stoichiometric mixtures of the corresponding elements in an argon atmosphere at a pressure of 10^{-4} mbar. Care was taken to avoid oxygen contamination. The samples were annealed afterward for 2 weeks at 1073 K in an evacuated quartz tube. After the annealing process, the samples were quenched in a mixture of ice and water to retain the desired structure. ZrNi$_2$Ga crystallizes in the cubic L2$_1$ Heusler structure (space group: $Fm\bar{3}m$), where the Wyckoff positions are 4a (0,0,0) for Zr atoms, 4b ($\frac{1}{2},\frac{1}{2},\frac{1}{2}$) for Ga atoms, and 8$c$ ($\frac{1}{4},\frac{1}{4},\frac{1}{4}$) for Ni atoms. The unit cell of the Heusler structure is displayed in Figure 8.1. The crystal structure of ZrNi$_2$Ga was investigated using powder x-ray diffraction (XRD). The measurements were carried out using a Siemens D5000 with monochromatized Cu K_α radiation. The electrical resistance of a bar-shaped sample was measured using a four-point probe technique. The magnetization measurements below a temperature of 4 K were performed in a superconducting quantum interference device (SQUID) (Quantum Design MPMS-XL-5). For higher temperatures, the magnetization was measured using a vibrating sample magnetometer (VSM) (VSM option of a Quantum Design PPMS). The measured samples had a spherical shape with a mass of approximately 20-120 mg. In order to study the diamagnetic shielding, the sample was initially cooled down to $T = 1.8$ K without applying any magnetic field, i.e., zero-field cooled (ZFC). Then a field of $\mu_0 H = 2.5$ mT was applied, and the sample magnetization was recorded with increasing temperature. To determine the Meissner effect (flux expulsion) the sample was subsequently cooled and its magnetization measured in the identical field, i.e., field cooled (FC). The field-dependent magnetization of ZrNi$_2$Ga was measured at a temperature of 2 K. Finally, the normal-state susceptibility was measured at $\mu_0 H = 2$ T in a temperature range from 1.8 K to 300 K. Specific-heat measurements were carried out at 0.35 K $< T <$ 4 K in magnetic fields of up to 5 T in a Quantum Design PPMS with a ^{3}He option.

8. Ni-based superconductor: Heusler compound ZrNi$_2$Ga

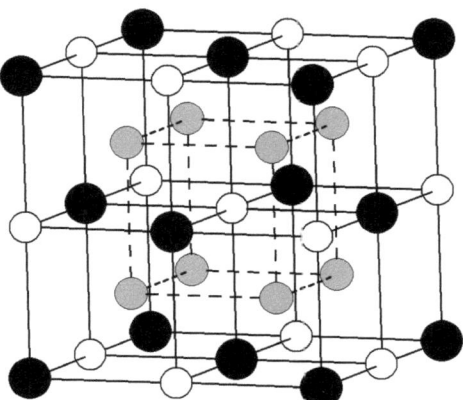

Figure 8.1: The cubic $L2_1$ Heusler structure of ZrNi$_2$Ga. Zr atoms are black, Ga atoms white, and Ni atoms are gray. The Zr atoms are placed on Wyckoff position $4a$ $(0,0,0)$, Ga on $4b$ $(\frac{1}{2},\frac{1}{2},\frac{1}{2})$ and Ni on $8c$ $(\frac{1}{4},\frac{1}{4},\frac{1}{4})$. The Ni atoms build up a simple-cubic sublattice. The centers of the cubes formed by Ni are alternately occupied by Zr and Ga atoms corresponding to a perfect 2^3 CsCl superstructure.

Ab initio calculations of the electronic and vibrational properties.

The electronic and vibrational properties were calculated through the use of WIEN2k (Ref. [99]) in combination with PHONON [80]. The electronic structure of ZrNi$_2$Ga was calculated by means of the full potential linearized augmented plane-wave (FLAPW) method as implemented in WIEN2k provided by Blaha, Schwartz, and co-workers [74, 79, 99]. The exchange-correlation functional was taken within the generalized gradient approximation (GGA) in the parameterization of Perdew, Burke and Enzerhof [75]. A $25 \times 25 \times 25$ point mesh was used as base for the integration in the cubic systems resulting in 455 k points in the irreducible wedge of the Brillouin zone. The energy convergence criterion was set to 10^{-5} Ry and simultaneously the criterion for charge convergence to $10^{-3} e^-$. The muffin-tin radii were set to 2.5 a_{0B} ($a_{0B} :=$ Bohr's radius) for the transition metals as well as the main group element. A volume optimization resulted in $a_{opt} = 6.14$ Å and a bulk modulus of $B = 156$ GPa for the relaxed structure. This value is slightly larger than the experimentally observed lattice parameter a_{exp} (see below). The results presented in the following are for the relaxed lattice parameter. No noticeable changes are observed in the calculations using a_{exp}.

8. Ni-based superconductor: Heusler compound ZrNi$_2$Ga

Figure 8.2 shows the results for the electronic structure from the *ab initio* calculations. Typical for Heusler compounds is the low-lying hybridization gap at energies between 7 eV and 5.6 eV below the Fermi energy. This gap emerges from the strong interaction of the $s - p$ states at the Ga atoms in O_h symmetry with the eight surrounding Ni atoms. It explains the structural stability of the compound.

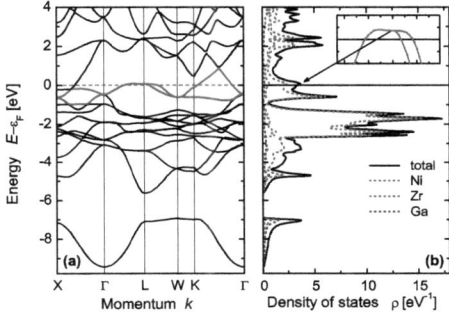

Figure 8.2: Electronic structure of ZrNi$_2$Ga. (a) displays the band structure and (b) the density of states. The inset in (b) shows the dispersion of the bands that cause the van Hove singularity at the L point on an enlarged scale.

More interesting are the bands close to the Fermi energy. In particular, the topmost valence band exhibits a van Hove singularity at the L point only 70 meV above ϵ_F. The result is a maximum of the density of states at the Fermi energy [see Figure 8.2(b)]. A closer inspection of those states reveals that the singularity at L is an S_2-type saddle point of the electronic structure with a twofold degeneracy. This degeneracy is removed along LK or LW. For both bands, two of the second derivatives $\left|\partial^2 E(k)/\partial k_i \partial k_j\right|_{k_e}$ of the dispersion $E(k)$ are > 0 and one is < 0 (Λ-direction) at $k_e = (1/4, 1/4, 1/4)$.

Figure 8.3 shows the calculated phonon dispersion and phonon density of states. The dispersion of the acoustic LA and TA_1 modes is degenerate in the fourfold Δ direction as well as along Λ. This degeneracy is removed at the K point and in the twofold Σ direction. Instabilities in the form of soft-phonon modes, as are observed for several magnetic Ni-based Heusler compounds [89, 100], do not occur in the phonon-dispersion relation of ZrNi$_2$Ga. This indicates the high structural stability of the compound compared to the Ni-based Heusler shape memory alloys (for example MnNi$_2$Ga).

The high density of phonons at energies of about 30 meV is due to the vibration of the rather heavy Zr atoms. These optical modes have no overlap with the remainder of the phonon spectrum and appear as Einstein frequencies. In a hybrid Einstein-Debye model, this corresponds to an Einstein temperature of $\Theta_E \approx 340$ K and a Debye temperature of $\Theta_D \approx 270$ K taken from the density maximum at the upper cutoff of the optical modes.

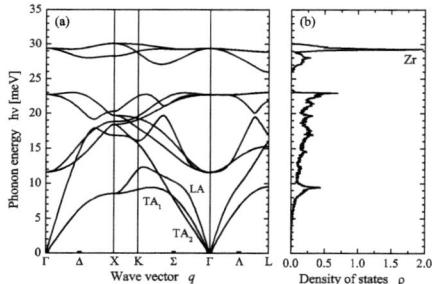

Figure 8.3: The calculated vibrational spectrum of ZrNi$_2$Ga. (a) displays the phonon dispersion and (b) the corresponding density of states.

Results and Discussion

Crystal structure and sample quality

The cubic L2$_1$ Heusler structure of ZrNi$_2$Ga was determined using XRD. Figure 8.4 shows the diffraction pattern for ZrNi$_2$Ga with the raw data above (black) and the difference between a calculated Rietveld refinement and the raw data below (gray). Within the experimental resolution of the diffractometer, no secondary phases were observed. The Rietveld refinement results in a cubic lattice parameter of $a = 6.098 \pm 0.003$ Å. The as-cast samples of ZrNi$_2$Ga were indistinguishable from the annealed ones in their XRD patterns; but magnetic, transport, and specific-heat measurements suggested an improved quality of the annealed samples. This improved quality of the annealed crystals was confirmed by resistivity measurements yielding a residual resistivity ratio of two, which is typical for polycrystalline Heusler compounds. The specific-heat and magnetization measurements reveal sharp superconducting transitions of $\Delta T_c/T_c \leq 0.03$. At low temperature, however, the measurements indicate small sample inhomogeneities or impurities, which are discussed in Section 8.

8. Ni-based superconductor: Heusler compound ZrNi$_2$Ga

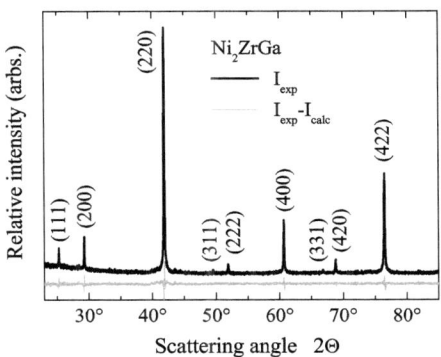

Figure 8.4: Powder x-ray diffraction of ZrNi$_2$Ga at 300 K (black). The difference curve (gray) shows the difference between the observed data and the Rietveld refinement.

Properties of the superconducting state

The superconducting transition of ZrNi$_2$Ga was observed in measurements of the electrical resistance. Figure 8.5 displays the temperature dependence of the resistance, which exhibits metallic behavior and a transition to superconductivity at $T_c = 2.87 \pm 0.03$ K.

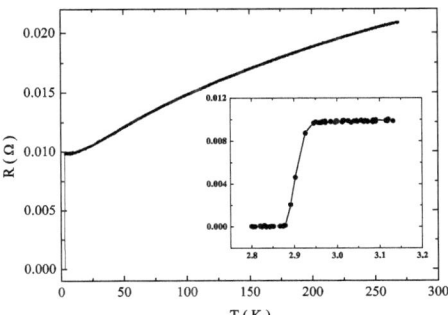

Figure 8.5: The resistance of ZrNi$_2$Ga as a function of temperature. The inset shows an enlargement of the superconduction transition at $T_c^{mid} = 2.87$ K.

Magnetization measurements using SQUID magnetometry were carried out to confirm bulk superconductivity in ZrNi$_2$Ga. The results of the magnetization measurements are given in Figure 8.6. The upper panel (a) shows the temperature dependent magnetization $M(T)$ of a nearly spherical sample in an external field of $\mu_0 H = 2.5$ mT. A sharp onset of superconductivity is observed in the ZFC curve at a temperature of $T_c = 2.80$ K. The sharpness of the transition indicates good sample quality. The resistive transition appears at a slightly higher temperature than that determined from the magnetization measurements. This is a well-known phenomenon: the resistive transition occurs when one percolation path through the sample becomes superconducting, whereas the magnetic transition requires a certain superconducting volume.

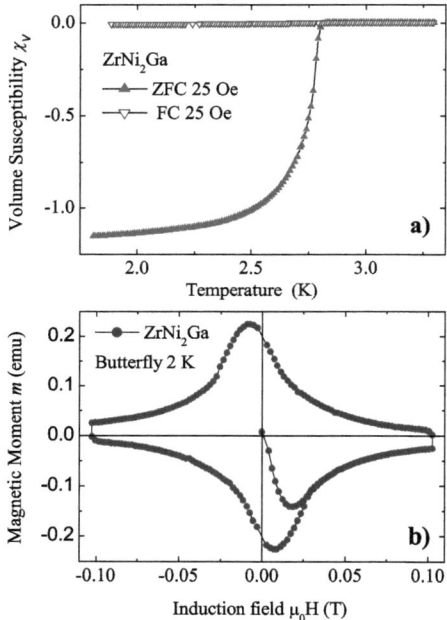

Figure 8.6: Magnetization measurements in the superconducting state of ZrNi$_2$Ga. (a) shows the temperature-dependent magnetization under ZFC and FC conditions. Panel (b) shows the field-dependent magnetization at a temperature of 2 K.

The ZFC curve demonstrates complete diamagnetic shielding. For the calculation of the magnetic volume susceptibility, we used the demagnetization factor of $\frac{1}{3}$ of a sphere. The deviation from the expected value of -1 (100% shielding) is ascribed to an imper-

8. Ni-based superconductor: Heusler compound ZrNi$_2$Ga

fect spherical shape of the sample and therefore an underestimated demagnetization factor. The FC curve represents the Meissner effect for superconducting ZrNi$_2$Ga. The large difference between the ZFC and the FC curves shows clearly that ZrNi$_2$Ga is a type-II superconductor and points to a weak Meissner effect due to strong flux pinning. Figure 8.6(b) shows a plot of the field-dependent magnetization (M-H curve). The magnetic field was varied from -100 mT to 100 mT at a constant temperature of 2 K. The $M(H)$ measurements exhibit the typical butterfly loop of an irreversible type-II superconductor with large hysteresis due to strong flux pinning. An accurate determination of the lower critical magnetic field H_{c1} at this temperature is nearly not possible because of the broadening of the $M(H)$ curves. A very rough estimation of H_{c1}, defined as the magnetic field where the initial slope interacts with the extrapolation curve of $(M_{up} + M_{down})/2$, yields $\mu_0 H_{c1}(T = 2\ \text{K})$ of approximately 16 mT compared to the upper critical field at $T = 2$ K of 0.62 T.

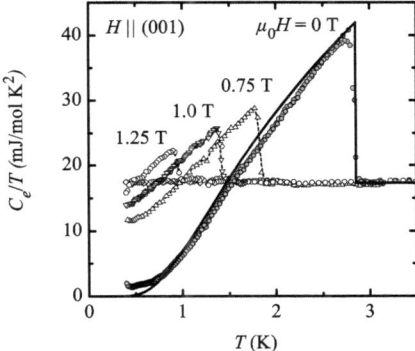

Figure 8.7: Electronic contribution to the specific heat of ZrNi$_2$Ga divided by temperature T at various magnetic fields. The continuous line represents the calculated behavior of a weak-coupling BCS superconductor at zero magnetic field.

Figure 8.7 shows the electronic contribution to the specific heat C_e of ZrNi$_2$Ga plotted as C_e/T vs T in various magnetic fields. The phonon contribution to the specific heat was subtracted as will be shown below. The main feature of C_e/T is the specific-heat jump ΔC_e at $T_c = 2.83$ K with a width of 0.1 K. The nearly perfect agreement between the differently determined T_c values together with the large ΔC_e confirms bulk superconductivity in ZrNi$_2$Ga. An analysis of the jump yields $\Delta C_e/\gamma_n T_c = 1.41$, which is in very good agreement with the weak-coupling BCS value of 1.43. Here γ_n denotes the normal-state Sommerfeld coefficient, which is discussed below. The energy gap is obtained from a plot of $C_e/\gamma T_c$ on a logarithmic scale versus T_c/T, as shown in Figure 8.8. A comparison with the BCS formula for C_e well below T_c

$$C_e/\gamma T_c = 8.5 \exp[-(0.82\Delta(0)/k_B T]$$

yields an energy gap $\Delta(0)$ of 0.434 meV for $T \to 0$ and $2\Delta(0)/k_B T_c = 3.53$, which is again in very good agreement with the weak-coupling BCS value.

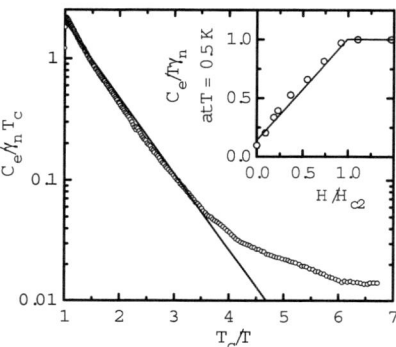

Figure 8.8: Electronic contribution to the specific heat of ZrNi$_2$Ga at zero field divided by $\gamma_n T_c/T$ vs. T_c/T. The inset shows C_e/T at $T = 0.5$ versus the magnetic field.

At lowest temperatures one can observe deviations from the expected behavior. As these deviations are sample dependent and clearly reduced in the annealed samples, we attribute them to the aforementioned sample imperfections. In a more detailed analysis we compared C_e at zero field with the calculated behavior of a BCS superconductor by using the approach of Padamsee et al. [101] and the temperature dependence of the gap $\Delta(T)$ of Mühlschlegel [102]. In this model, C_e is estimated for a system of independent fermion quasiparticles with

$$\frac{S}{\gamma_n T_c} = -\frac{6}{\pi^2}\frac{\Delta(0)}{k_B T_c}\int_0^\infty [f \ln f + (1-f)\ln(1-f)]dy,$$

$$\frac{C_e}{\gamma_n T_c} = t\frac{\partial(S/\gamma_n T_c)}{\partial t}$$

where

$$f = [\exp(\sqrt{\epsilon^2 + \Delta^2(t)})/k_B T + 1],\ t = T/T_c,\ y = \epsilon/\Delta_0.$$

The only free parameter, the ratio $2\Delta(0)/k_B T_c$, was set to 3.53. Indeed, the specific

8. Ni-based superconductor: Heusler compound ZrNi$_2$Ga

heat can overall be rather well described by the weak-coupling BCS theory, as can be seen in Figure 8.7. To study the influence of the magnetic field we plot C_e/T at a constant temperature of 0.5 K vs the H/H_{c2} in the inset of Figure 8.8. The linear increase of C_e/T with H corresponds to an isotropic gap, as expected for a cubic BCS superconductor.

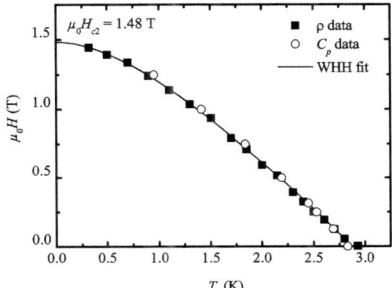

Figure 8.9: Temperature dependence of the upper critical field H_{c2} of ZrNi$_2$Ga. Shown is a summary of the resistance and specific-heat measurements. The continuous line represents a calculation of the WHH model with $\alpha = 0$ and $\lambda_{so} = 0$, which is identical to a finite α and $\lambda_{so} \to \infty$.

Further $R(T)$ measurements in various magnetic fields were performed to determine the upper critical field H_{c2} of ZrNi$_2$Ga. In Figure 8.9 the data are summarized together with those of the specific-heat measurements. $H_{c2}(T)$ was theoretically derived by Werthamer, Helfand, and Hohenberg (WHH) (Ref. [103]) in the limit of short electronic mean free path (dirty limit), including, apart from the usual orbital pair breaking, the effects of Pauli-spin paramagnetism and spin-orbit scattering. The model has two adjustable parameters: the Maki parameter α, which represents the limitation of H_{c2} by the Pauli paramagnetism, and the spin-orbit scattering constant λ_{so}. α can be determined from the initial slope of the upper critical field

$$\alpha = -0.53 \cdot \mu_0 \, dH_{c2}/dT|_{T=T_c} \, (\mu_0 H \text{ in T}),$$

or via the Sommerfeld coefficient γ_n and the residual resistivity ρ_0 with:

$$\alpha = 2e^2 \hbar \gamma_n \rho_0 / (2\pi^2 m k_B^2),$$

where m and e are the free electron mass and charge, respectively. From the data we extract $\mu_0 \, dH_{c2}/dT|_{T=T_c} = -0.75$ T/K and $\alpha = 0.4$. With $\lambda_{so} \to \infty$, the curve estimated by the WHH model follows the data points very closely, as is seen in Figure 8.9. As the spin-orbit scattering counteracts the effect of the Pauli paramagnetism, this is equal to $\alpha = 0$ and $\lambda_{so} = 0$, representing the upper bound of H_{c2} where pair breaking is only induced by orbital fields. Consequently, the temperature dependence of H_{c2} can either be explained by Pauli paramagnetism with an extremely strong spin-orbit scattering or with a dominating orbital field effect. The critical field due to the Pauli term alone is $\mu_0 H_p(0) = \mu_0 \Delta(0)/\sqrt{2}\mu_B = 1.84 T_c = 5.24$ T, which is much higher than H_{c2} in the absence of Pauli paramagnetism $\mu_0 H_{c2}^*(0) = -0.69 \cdot \mu_0 \, dH_{c2}/dT|_{T=T_c} = 1.48$ T. Hence pair breaking in ZrNi$_2$Ga is most probably only caused by orbital fields [104]. This is in contrast to other Ni-based Heusler superconductors like NbNi$_2$Ga and NbNi$_2$Sn where $H_{c2}^*(0)$ is clearly larger than the measured critical fields and therefore the Pauli paramagnetic effect has to be considered (see Table 8.1).

Table 8.1: Comparison of nickel-based paramagnetic and superconducting Heusler compounds. Sommerfeld coefficient γ_n, Debye temperature Θ_D, superconducting transition temperature T_c, orbital limit of the upper critical field $\mu_0 H_{c2}^*(0) = -0.69 \cdot \mu_0 \, dH_{c2}/dT|_{T=T_c}$, and critical field $H_{c2}(0)$ extrapolated from low-temperature measurements.

	γ_n (mJ/mol K^2)	Θ_D (K)	T_c (K)	$\mu_0 H_{c2}^*$ (T)	$\mu_0 H_{c2}$ (T)
ZrNi$_2$Ga	17.3	300[1]/270[2]	2.85	1.48	1.48
TiNi$_2$Al	13.37[3]	411[3]	-	-	-
TiNi$_2$Sn	6.86[4]	290[4]	-	-	-
ZrNi$_2$Al	13.67[3]	276[3]	-	-	-
ZrNi$_2$Sn	8.36[4]	318[4]	-	-	-
HfNi$_2$Al	10.85[3]	287[3]	-	-	-
HfNi$_2$Sn	6.37[4]	280[4]	-	-	-
VNi$_2$Al	14.17[3]	358[3]	-	-	-
NbNi$_2$Al	8.00[5],10.95[3]	280[5],300[3]	2.15[5]	0.96[6]	> 0.70[5]
NbNi$_2$Ga	6.50[5]	240[5]	1.54[5]	0.67[6]	~ 0.60[5]
NbNi$_2$Sn	4.0[5],5.15[3]	206[5],208[3]	2.90[5],3.40[3]	0.78[6]	~ 0.63[5]
TaNi$_2$Al	10.01[3]	299[3]	-	-	-

The thermodynamic critical field was calculated from the difference between the free energy of the superconducting and normal states as follows:

$$\mu_0 H_c = \left[2\mu_0 \int_{T_c}^{T} \int_{T_c}^{T} (C_e/T'' - \gamma_n) dT'' dT' \right]^{\frac{1}{2}}.$$

---

[1] Measured value in this work
[2] Calculated value in this work
[3] Reference [95]
[4] Reference [105]
[5] Reference [106]
[6] Calculated with the initial slope dH_{c2}/dT from Ref. [95]

8. Ni-based superconductor: Heusler compound ZrNi$_2$Ga

The value of $\mu_0 H_c = 44.6$ mT is obtained. From the upper and thermodynamic critical fields one can estimate the Ginzburg-Landau parameter κ_{GL}, which is the ratio of the spatial variation length of the local magnetic field λ_{GL} and the coherence length ξ_{GL}: $\kappa_{GL} = H_{c2}(\sqrt{2}H_c) = \lambda_{GL}/\xi_{GL} = 23.5$. The isotropic Ginsburg-Landau-Abrikosov-Gor'kov theory leads to the values of $\xi_{GL} = \sqrt{\Phi_0/2\pi\mu_0 H_{c2}} = 15$ nm and $\lambda_{GL} = 350$ nm (Φ_0 is the fluxoid quantum $h/2e$).

Obviously, ZrNi$_2$Ga is a conventional, weakly coupled, fully gapped type-II superconductor that is best described in terms of weak-coupling BCS superconductivity. If a phonon-mediated pairing mechanism is assumed, we can determine the dimensionless electron-phonon coupling constant λ by using the McMillan relation:[107]

$$T_c = \frac{\Theta_D}{1.45} \exp\left[\frac{-1.04(1+\lambda)}{\lambda - \mu_c^*(1+0.62\lambda)}\right].$$

If the Coulomb coupling constant μ_c^* is set to its usual value of 0.13 and Θ_D to our measured value of 300 K we get $\lambda = 0.551$, which is in good accordance with other superconducting Heusler compounds [95].

Normal-state properties

Now we turn to a characterization of the normal-state properties. When superconductivity is suppressed in a magnetic field of $H > H_{c2}$, the Sommerfeld coefficient γ_n and the Debye temperature Θ_D can be extracted from the low-temperature behavior of the specific heat, $C = \gamma_n T + \frac{12}{5}\pi^4 R n \theta_D^{-1} T^3$ where R is the gas constant and n is the number of atoms per formula unit (= 4 in the case of Heusler compounds). The extracted Debye temperature $\Theta_D = 300$ K agrees very well with the calculated value of 270 K and is in the typical Θ_D range of other Heusler compounds (see Table 8.1).

Likewise in accordance to our electronic structure calculations, the high density of states leads to a strongly enhanced Sommerfeld coefficient of $\gamma_n = \frac{\pi^2}{3}k_B^2 N(\epsilon_F) = 17.3$ mJ/mol K^2. In fact, γ_n is one of the highest values for paramagnetic Ni-based Heusler compounds (see Table 8.1). As already stated by Boff et al. [106], the maximum of γ_n in the isoelectronic sequence A = Ti, Zr, Hf of ANi$_2 C$ (C = Al, Sn) is found for Zr and in the sequence A = V, Nb, Ta for V. As the electronic structure of all these compounds is quite similar, and consequently a rigid-band model may be applicable, the Fermi level can be shifted through the appropriate choice of A to a maximum of $N(\epsilon_F)$ [106, 108, 109]. This behavior and the comparatively large γ_n of ZrNi$_2$Ga confirm the van Hove scenario.

The measured magnetic susceptibility $\chi(T)$ as shown in Figure 8.10 is nearly independent of T, which is indicative of a predominantly Pauli-like susceptibility. No sign of magnetic order can be found down to $T = 1.8$ K. Even more, the low-temperature specific-heat measurements demonstrate clearly that apart from the superconductivity no other phase transitions occur down to temperatures of 0.35 K. The enhanced susceptibility corresponds to the high density of states seen in γ_n value as evidenced by the Wilson ratio $R = (\chi/\gamma_n) \cdot \pi^2 k_B^2 / 3\mu_0 \mu_{eff}^2 = 0.97$, where we have set $\mu_{eff}^2 = g^2 \mu_B^2 J(J+1)$ to its free-electron values, i.e., the Landé factor $g = 2$ and the total angular momentum $J = \frac{1}{2}$. The resulting Wilson ratio is close to that for independent electrons ($R = 1$).

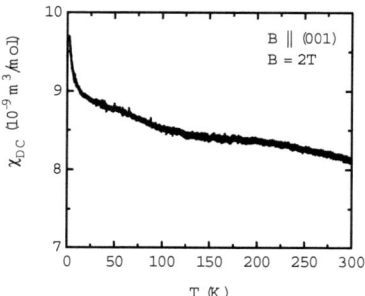

Figure 8.10: Susceptibility $\chi_{DC} = M/H$ of ZrNi$_2$Ga in a magnetic field of $\mu_0 H = 2$ T$> \mu_0 H_{c2}$. The susceptibility of the normal state shows Pauli-type behavior without any indications of magnetic order. At low temperatures there is a small Curie-Weiss-type upturn, which may be attributed to sample inhomogeneities or impurities.

Below about 10 K, a Curie-Weiss-type increase of χ is observed for all samples. A fit of a Curie-Weiss law to the data yields a Weiss temperature of -3.3 K and an effective moment of 0.06 μ_B/f.u. (assuming $s = 1/2$). This Curie-Weiss-type behavior is sample dependent and can again be attributed to a small amount of magnetic impurities. It is, however, supervising that no appreciable pair breaking is observed as evidenced by the validity of the BCS law of corresponding states $2\Delta(0) = 3.53 k_B T_c$.

Finally, we want to discuss the influence of the increased DOS on the superconducting properties of ZrNi$_2$Ga. Although ZrNi$_2$Ga exhibits an enhanced γ_n compared to the value 5.15 mJ/mol K^2 of NbNi$_2$Sn, both compounds have nearly the same transition temperature. Obviously, the simple relationship between $N(\epsilon_F)$ and T_c does not hold. Table 8.1 demonstrates, likewise, that the upper critical field H_{c2} and the orbital limit H_{c2}^* apparently do not depend on the density of states in these materials.

Electron doping

The influence of the increased DOS on the superconducting properties is investigated from another point of view, which refers only to ZrNi$_2$Ga and the van Hove singularity in this compound. The Fermi level can be shifted with an appropriate choice of the element A within the Heusler formula unit AB_2C, and the van Hove scenario yields a maximum T$_c$ when the van Hove singularity coincides with ϵ_F. According to the electronic structure calculations, electron doping of ZrNi$_2$Ga should lead to this desired coincidence. Therefore, we doped ZrNi$_2$Ga with electrons in the A position by substituting Zr with distinct amounts of Nb. The alloys Zr$_{1-x}$Nb$_x$Ni$_2$Ga with x=0.15, 0.3, 0.5, and 0.7 were prepared according to Section 8.

The crystal structures of the alloys were determined using a Siemens D8 Advance diffractometer with Mo K_α radiation. All alloys were found to crystallize in the Heusler structure (space group: $Fm\bar{3}m$). The atomic radius of Nb is smaller than the one of Zr, and thus a decrease of the lattice parameter is expected upon substituting Zr with Nb. In fact, this effect was observed (Figure 8.11). No impurity phases were detected in all alloys except of Zr$_{0.3}$Nb$_{0.7}$Ni$_2$Ga. The small difference between the lattice parameters of Zr$_{0.3}$Nb$_{0.7}$Ni$_2$Ga and Zr$_{0.5}$Nb$_{0.5}$Ni$_2$Ga supports that a saturation of Nb in the lattice of Zr$_{1-x}$Nb$_x$Ni$_2$Ga is reached for a value of 0.5≤x≤0.7. Increasing the Nb concentration above the saturation limit leads to segregation of impurities. One of them was identified as elemental Zr.

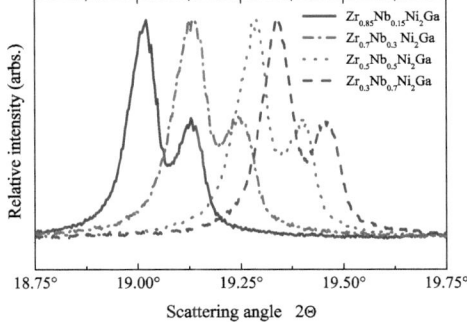

Figure 8.11: Powder x-ray diffraction of the alloys Zr$_{1-x}$Nb$_x$Ni$_2$Ga at 300 K. Shown is the region around the (220) reflection, which determines the cubic lattice parameter. The signals are splitted in Mo $K_{\alpha 1}$ and Mo $K_{\alpha 2}$ peaks.

The superconducting transitions of the alloys were analyzed in magnetization measurements using SQUID magnetomety as described in Section 8. Figure 8.12 shows the

ZFC curves of the alloys $Zr_{0.85}Nb_{0.15}Ni_2Ga$, $Zr_{0.7}Nb_{0.3}Ni_2Ga$, and $Zr_{0.5}Nb_{0.5}Ni_2Ga$. $Zr_{0.3}Nb_{0.7}Ni_2Ga$ did not show a superconducting transition down to 1.8 K. This is not surprising because of the impurities, which were detected from XRD in this alloy. The other alloys show a trend of decreasing T_c with increasing Nb concentration as summarized in Table 8.2. The expected behavior, a maximum T_c at a certain Nb concentration, was thus not found. According to Anderson's theorem, nonmagnetic scattering should not suppress superconductivity by itself [93]. Ma and Lee [94] showed, however, that enhanced disorder leads to spatial fluctuations of Δ and eventually to the suppression of the superconducting state. Obviously, even the lowest Zr:Nb ratio of 85:15 is sufficient to provoke a high degree of disorder in $Zr_{0.85}Nb_{0.15}Ni_2Ga$, which leads to reduction of T_c and suppression of superconductivity. Furthermore, the alloys cannot be described accurately with the rigid-band model. Doping with Nb leads to smearing out of the states and to a broadening of the van Hove singularity.

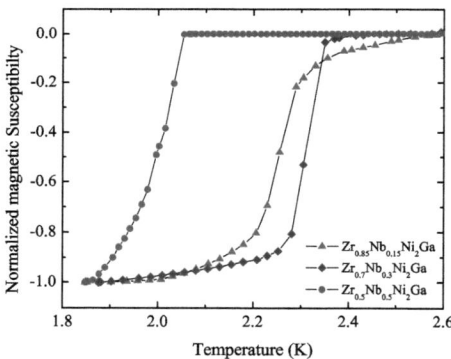

Figure 8.12: Superconducting transitions of the alloys $Zr_{1-x}Nb_xNi_2Ga$ under ZFC conditions. The measurements were performed with magnetic fields of $\mu_0 H = 2.5$ mT.

Conclusions

Starting with electronic structure calculations, the Heusler compound $ZrNi_2Ga$ was predicted to have an enhanced density of states at the Fermi energy $N(\epsilon_F)$ due to a van Hove singularity close to ϵ_F. According to the BCS model, $ZrNi_2Ga$ was therefore expected to be an appropriate candidate for superconductivity with a comparatively high superconducting transition temperature.

Table 8.2: Properties of the alloys $Zr_{1-x}Nb_xNi_2Ga$ compared to $ZrNi_2Ga$. a is the measured lattice parameter and T_c is the critical temperature from the ZFC curves in the magnetization measurements.

Compound/alloy	a (Å)	T_c (K)
$ZrNi_2Ga$	6.098	2.8
$Zr_{0.85}Nb_{0.15}Ni_2Ga$	6.074	2.4
$Zr_{0.7}Nb_{0.3}Ni_2Ga$	6.037	2.3
$Zr_{0.5}Nb_{0.5}Ni_2Ga$	5.990	2.0
$Zr_{0.3}Nb_{0.7}Ni_2Ga$	5.972	-

The predicted superconducting transition was found at $T_c = 2.87$ K. Specific-heat and magnetization measurements proved bulk superconductivity in this material and demonstrate that $ZrNi_2Ga$ is a conventional, weakly coupled BCS type-II superconductor. The electronic specific heat of the normal state shows a clearly enhanced Sommerfeld coefficient γ_n, which supports the van Hove scenario. In the temperature range 0.35 K $< T <$ 300 K, no sign of magnetic order is found. Apparently, the high $N(\epsilon_F)$ is not sufficient to satisfy the Stoner criterion. The normal-state susceptibility is described best by an increased Pauli paramagnetism, corresponding to an enhanced $N(\epsilon_F)$. Despite the presence of magnetic impurities, which would suppress the energy gap by pair breaking, the BCS law of corresponding states holds. This point deserves further investigations.

Acknowledgements

This work is funded by the DFG in Collaborative Research Center *"Condensed Matter Systems with Variable Many-Body Interactions"* (Transregio SFB/TRR 49). The work at Princeton was supported by the US Department of Energy division of Basic Energy Sciences, grant DE-FG02-98ER45706. The authors would like to thank Gerhard Jakob for many suggestions and for fruitful discussions.

9 Rational design of a novel noncentrosymmetric superconductor

The text of this chapter is identical with the following manuscript, which is intended to be submitted for publication:
J. Winterlik, G. H. Fecher, C. Felser
To be submitted to Angew. Chem. Int. Ed. (2009)

Ternary intermetallic Heusler compounds with the stoichiometric composition A_2BC, where A and B are transition metals, and C is a main group element, have come to attention with the archetype Cu_2MnAl [78], a remarkable ferromagnet despite the absence of any ferromagnetic element. Heusler compounds crystallize in the cubic space group no. 225 ($Fm\bar{3}m$). Nowadays Heusler compounds are mainly associated with the research area of spintronics applications, *i. e.* as half-metals, where due to the exchange splitting of the d-electron states, only electrons of one spin direction exhibit a density of states at the Fermi energy $n(\epsilon_F)$ [62, 63].

Up to the present, only a few Heusler compounds have been found to exhibit superconductivity. The first Heusler superconductors were reported in 1982 [66]. Among them, Pd_2YSn exhibits the highest yet recorded critical temperature Tc of 4.9 K [67]. Pd_2YbSn and Pd_2ErSn, both with rare earth metals at the B position, exhibit coexistence of superconductivity and antiferromagnetic order [68, 69]. Very recently we employed electronic structure calculations in order to explore Heusler compounds as candidates for superconductivity and found several Pd- and Ni-based Heusler compounds, which become superconducting in the low temperature region [11–13].

The recipe we used in these approaches is based on the van Hove scenario [14]. The Bardeen-Cooper-Schrieffer (BCS) theory for superconductivity states that the transition temperature of a BCS superconductor increases exponentially with an increasing density of states at the Fermi energy $n(\epsilon_F)$ provided that Debye temperature Θ_D and Cooper-pairing interaction are independent $n(\epsilon_F)$. Saddle points in the energy dispersion curves of the electronic structure of solids lead to maxima in the density of states (DOS), the van Hove singularities. The coincidence of a saddle point and thus a van Hove singularity with the Fermi energy in a compound causes a high density of states at ϵ_F and is therefore a good precondition for superconductivity. Assuming this coincidence in the van Hove scenario, Θ_D can be substituted by the Fermi temperature Θ_F, which allows higher transition temperatures compared to the BCS theory [110, 111]. The van Hove scenario was used to explain the unusually high transition temperatures of the intermetallic A15 superconductors [98] and of the cuprate superconductors [112]. The energy dispersion curve of $YBa_2Cu_3O_7$ (YBCO) exhibits a saddle point at the X-point in the near of the Fermi edge. This feature, which was verified experimentally using angle-resolved photoemission [113], was used for an explanation of the high transition temperature of YBCO according to the van Hove scenario [114, 115]. We have figured out using electronic structure calculations that certain Heusler compounds with 27 electrons exhibit saddle points at the L-points in the energy dispersion curves in the near of the Fermi edge. Superconductivity of these compounds was verified according to the van Hove scenario [11–13]. The idea of this work was the design of a novel intermetallic superconductor without inversion symmetry. Noncentrosymmetric superconductors are of utmost importance with respect to understanding superconducting

9. Rational design of a novel noncentrosymmetric superconductor

gap symmetries and pairing mechanisms. The major concern of these superconductors is their pairing state when the inversion center is lacking. Parity is no longer symmetry of the Cooper pairs. The pairing state should therefore be a mixture of singlet and triplet Cooper pairs. The discovery of superconductivity in noncentrosymmetric systems such as the heavy fermion superconductor $CePt_3Si$ [116] has raised the interest in elucidating the influence of the lacking inversion symmetry on the superconducting properties of a compound. The most attention has been paid to strongly correlated noncentrosymmetric heavy fermion superconductors such as $CePt_3Si$, $CeRhSi_3$ [117], $CeIrSi_3$ [118], and UIr [119] because of their unconventional behavior. Superconductivity without inversion symmetry has, however, also been found in more basic systems such as the transition metal compounds Y_2C_3 [120], Li_2Pd_3B [121], Li_2Pt_3B [122], $Mg_{10}Ir_{19}B_{16}$ [123], and KOs_2O_6 [124]. The anisotropic spin-orbit (SO) interaction in noncentrosymmetric compounds lifts spin degeneracy and splits the energy bands. The magnitude of SO coupling has thus an important influence on the superconducting gap symmetry and the Cooper pairing mechanism [125]. Li_2Pd_3B is a spin-singlet superconductor. Substitution of Pd by Pt and thus increasing the SO interaction leads to a nodal spin-triplet superconducting current in Li_2Pt_3B [126].

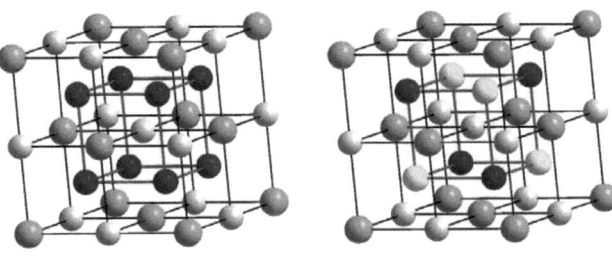

a) Heusler structure   b) LiMgPdSn structure

Figure 9.1: The Heusler structure in comparison to the LiMgPdSn structure. Both structures consist of four interpenetrating fcc lattices. In the Heusler structure, two of them are occupied by the same atom (Wyckoff position 8c).

The cubic LiMgPdSn structure (space group no. 216, $F\bar{4}3m$) is related very closely to the Heusler structure except for the lack of inversion symmetry as depicted in Figure 9.1. The electronic structures of Heusler compounds and compounds of the noncentrosymmetric LiMgPdSn structure are quite similar in many cases. We therefore applied the same design criterion as in the case of the Heusler superconductors and looked for noncentrosymmetric superconductor candidates that follow the van Hove scenario as described above for Heusler compounds, namely 27 electron compounds with van Hove singularities at the L-point. Using this method we identified PdAuScAl as a prospective candidate for noncentrosymmetric superconductivity. Successful preparation of the compound in the correct structure would also have additional advantages: Spin-orbit coupling should play an important role due to the heavy atoms Au and Pd. Disorder, which is a quite common feature in Heusler compounds, could be easily detected using x-ray diffraction because of the largely differing atomic scattering factors of Al, Sc, Pd, and Au, which are all located in different periods of the periodic table of elements. Calculational details, synthesis, investigation of superconducting properties and the structural characterization of PdAuScAl are described in the following.

The electronic structure of PdAuScAl was calculated using the full potential linearized augmented plane wave method as implemented in Wien2k [74].] The generalized gradient approximation was used for the exchange-correlation functional using the parameterizations of Perdew, Burke, and Enzerhof [75] or Engel and Vosko [127]. The band structure was calculated for the experimental lattice parameters. The equilibrium lattice parameters found from a volume optimization are slightly larger but leave the results unchanged. Due to the high Z of Au, the calculations were also performed including spin-orbit interaction. Figure 9.2 displays the calculated electronic structure of PdAuScAl. The splitting of the bands, caused by the spin orbit interaction, is clearly visible. The van Hove singularity at the L-point is found to be very close to the Fermi edge (approximately 0.35 eV above ϵ_F). The compound is therefore a convenient candidate for superconductivity according to the van Hove scenario. Coincidence of the maximum in the density of states with ϵ_F could be achieved with electron doping.

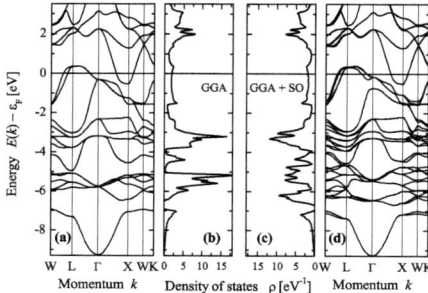

Figure 9.2: Electronic structure of PdAuScAl. Shown are the band structure (a, d) and the density of states (b, c). (a, b) are calculated without and (c, d) with inclusion of spin orbit interaction.

We have prepared polycrystalline ingots of PdAuScAl by repeated arc melting of stoichiometric mixtures of the corresponding elements in an argon atmosphere. Care was taken to avoid oxygen contamination. For a comparison of sample purity, some of the samples were annealed afterward for 2 weeks at 873 and 1073 K, respectively in evacuated quartz tubes. After the annealing process, the samples were quenched in a mixture of ice and water to retain the desired structure.

To account for superconductivity, the resistivity of PdAuScAl was measured using samples with polished surfaces, which are applied with electrical contacts according to the four-point probe technique. Figure 9.3 displays the resistance of PdAuScAl as a function of temperature at five different magnetic fields. As expected, a superconducting transition was found to take place at approximately 3.0 K in the field-free case. With increasing magnetic fields T_c is shifted to lower temperatures ($T_c = 2.3$ K for a magnetic induction field of 0.2 T). For a magnetic induction field of 0.5 T, no superconducting transition was found within the temperature range of the measurements (down to 1.8 K). The characteristic of the resistivity is metallic at high temperatures as is seen in the inset. The residual resistivity ratio of approximately 1.5 is a common value for intermetallic bulk samples and is comparable to values known from the class of Heusler compounds.

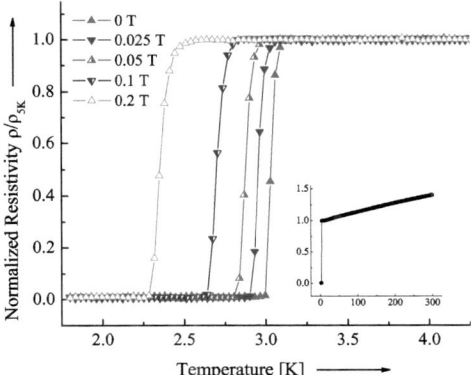

Figure 9.3: Resistivity of PdAuScAl. Shown is the temperature dependence of the resistivity in the region of T_c at five different magnetic fields. The inset shows the resistivity of PdAuScAl for an expanded temperature range.

Magnetization measurements using SQUID magnetometry were carried out to investigate the diamagnetic shielding and the Meissner effect of PdAuScAl bulk samples. The results of the magnetization measurements are given in Figure 9.4. The upper panel a) shows the temperature-dependent magnetic volume susceptibility $\chi_V(T)$ of PdAuScAl in an external induction field of 2.5 mT. For the calculation of $\chi_V(T)$, a demagnetization factor of 1/3 for a spherical sample geometry was incorporated. The ZFC curve shows the diamagnetic shielding for PdAuScAl. The onset of the superconducting transition is found to take place at a critical temperature of $T_c = 2.9$ K.

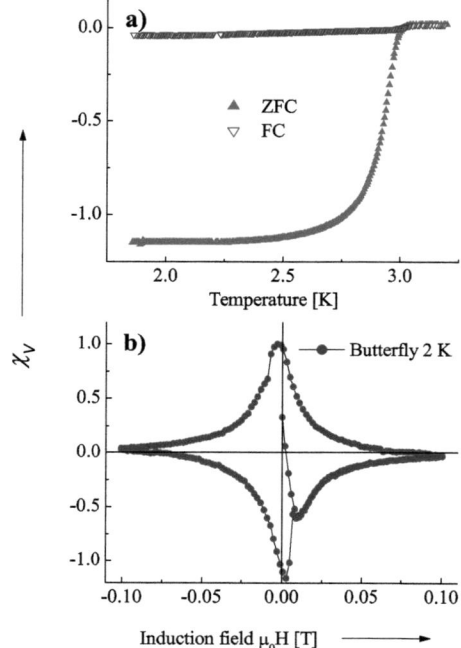

Figure 9.4: Diamagnetic shielding and Meissner effect of PdAuScAl. The temperature dependence of the susceptibility in a magnetic induction field of 2.5 mT under zero-field cooled (ZFC) and field cooled (FC) conditions is shown in a). The butterfly loop for PdAuScAl at 2 K is shown in b). The susceptibility values in b) are normalized by the maximum value.

The sharpness of the transition indicates good sample purity and quality. The incorporated demagnetization factor provides bulk superconductivity of roughly 115% for the sample. However, the factor of 1/3 may be too low because the sample does not exhibit the exact geometry of a perfect spherical body. The FC curve shows the Meissner effect for PdAuScAl. The large difference in magnitude between the ZFC and the FC measurements provides evidence that PdAuScAl is a type II superconductor and

9. Rational design of a novel noncentrosymmetric superconductor

points on a comparatively weak Meissner effect. This fact is attributed to strong flux pinning in the bulk material. The lower panel b) shows the field-dependent magnetization (butterfly loop) of PdAuScAl between -100 mT to 100 mT at a temperature of 2 K. An accurate determination of the critical magnetic field Hc1 at this temperature is impossible because of a broadening of the magnetization maximum. In addition, flux pinning causes a more symmetric shape of the curve. It is evident that the prediction of the van Hove scenario was verified for the novel superconductor PdAuScAl. The question of symmetry and the pairing mechanism are, however, yet unsolved. The crystal structure of PdAuScAl has to be analyzed before a statement can be made about these issues. The structure of PdAuScAl was determined using x-ray powder diffraction (XRD).

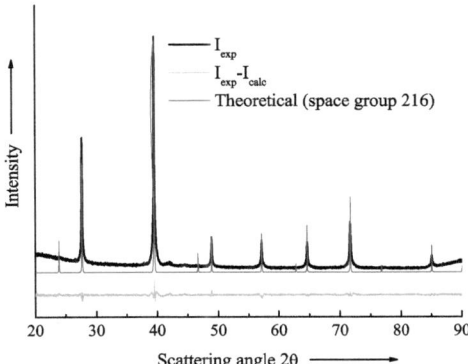

Figure 9.5: Powder x-ray diffraction of PdAuScAl at 300 K. The observed diffraction data (black) is shown above. The difference curve (grey) shows the difference between the observed data and the Rietveld refinement. The red pattern corresponds to a theoretical structure of a PdAuScAl compound with complete order among the 4c and the 4d sites.

Figure 9.5 shows the diffraction pattern of PdAuScAl. The cubic Heusler-type structure is easily recognized. The pattern indicates excellent phase purity of the samples. A buckling in the vicinity of the (220) reflection points on a small amount of an impurity, which is not yet identified. However, the XRD pattern of PdAuScAl definitely reveals that the compound does not crystallize in the LiMgPdSn structure. A detailed analysis of the (111), (311), (331), and (511) reflections shows that there is a high degree of disorder among the 4c and 4d sites in the cubic lattice (Pd and Au), which leads too a large decrease of the aforementioned reflections. The disorder creates a centre of symmetry and leads to a change of symmetry to $Fm\bar{3}m$. The Rietveld refinement was performed to an R-value of 3.2 and confirms the disorder among Pd and Au. A lattice parameter of $a = 6.442$ Å was derived from the refinement for PdAuScAl. Anti-site disorder is a common feature in Heusler compounds, and in many cases its amount can be reduced by annealing procedures. We therefore have annealed samples at 873 and 1073 K. These temperatures were useful for improving the order in the recently reported Pd-based Heusler superconductors [11, 13]. The XRD powder patterns of both samples, however, did not show any visible improvements with respect to the disorder between Pd and Au.

In conclusion, as we have predicted, the van Hove scenario works for the novel superconductor PdAuScAl. The compound is, however, not a noncentrosymmetric superconductor because disorder between Pd and Au creates an artificial center of inversion. This fact reveals the question, whether this artificial center of inversion and thus the disorder among the Pd and Au sites may be a necessary condition for superconductivity to occur in PdAuScAl. It may be deduced that the compound "wants to" escape the noncentrosymmetry to settle at an energetic minimum with the disordered structure in the conventional superconducting state. This question cannot yet be answered because the gap symmetry and the pairing mechanism of PdAuScAl require further experimental investigation. First measurements of the specific heat indicate that, besides the superconducting transition, the compound shows another second-order phase transition at approximately 2.3 K as depicted in Figure 9.6. The origin of this transition is not yet explained.

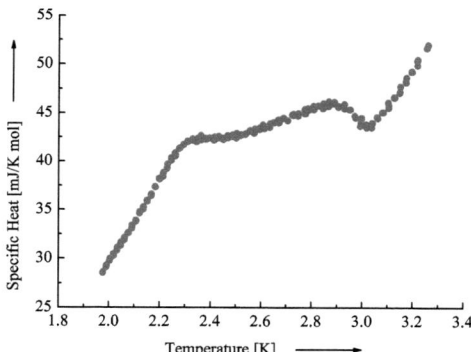

Figure 9.6: Powder x-ray diffraction of PdAuScAl at 300 K. The observed diffraction data (black) is shown above. The difference curve (grey) shows the difference between the observed data and the Rietveld refinement. The red pattern corresponds to a theoretical structure of a PdAuScAl compound with complete order among the 4c and the 4d sites.

Despite the occurrence of inversion symmetry and the fact that PdAuScAl behaves like a Heusler compound we were successful in the rational design of a novel superconductor. Further work has to be done to fully explore this interesting compound, primarily detailed investigations of the crystal structure, and the origin of the second low temperature phase transition. Development and application of working recipes such as the van Hove scenario can contribute fundamental findings for the elementary understanding of superconductivity.

Experimental Section

The transport properties (resistivity and specific heat) of the samples were analyzed using a Physical Property Measurement System (PPMS, Quantum Design, Model 6000). The magnetic properties were measured using a SQUID, Quantum Design, MPMS-XL-5. The XRD measurements were carried out using a Siemens D5000 with monochromatized Cu K_α radiation.

Acknowledgements

This work has been funded by the DFG [German Research Foundation] through the *"Condensed Matter Systems with Variable Many-Body Interactions"* Collaborative Research Center (Transregio SFB/TRR 49).

10 Structural, electronic, and magnetic properties of tetragonal $Mn_{3-x}Ga$: Experiments and first-principles calculations

The text of this chapter is identical with the following publication:
J. Winterlik, B. Balke, G. H. Fecher, C. Felser, M. C. M. Alves, F. Bernardi and J. Morais
Phys. Rev. B **77**, 054406 (2008).

Abstract

This work reports on the electronic, magnetic, and structural properties of the binary intermetallic compounds $Mn_{3-x}Ga$. The tetragonal DO_{22} phase of the $Mn_{3-x}Ga$ series, with x varying from 0 to 1.0 in steps of $x = 0.1$, was successfully synthesized and investigated. It was found that all these materials are hard magnetic, with energy products ranging from 10.1 kJm^{-3} for low Mn content ($x \to 1$) to 61.6 kJm^{-3} for high Mn content ($x \to 0$). With decreasing Mn content, the average saturation magnetization per atom increases from $0.26\mu_B$ for Mn_3Ga to $0.47\mu_B$ for Mn_2Ga. The increase in the saturation magnetization as the Mn content is reduced indicates a ferrimagnetic order with partially compensating moments of the two different Mn atoms on the two crystallographically different sites of the DO_{22} structure. This type of magnetic order is supported by *ab initio* calculations of the electronic structure that predict a nearly half-metallic ferrimagnet with the highest spin polarization of 88% at the Fermi energy for Mn_3Ga. The Curie temperature of the compounds is restricted to approximately 770 K because of a structural phase transition to the hexagonal DO_{19} phase. Thermal irreversibilities between zero-field-cooled and field-cooled measurements suggest that the $Mn_{3-x}Ga$ series belongs to the class of magnetically frustrated ferrimagnets. The most pronounced magnetic anomaly is found for Mn_3Ga.

Introduction

Several binary alloys Mn_3Z are known that contain manganese together with a main group element Z. Those compounds exhibit various different structures and magnetic properties. Cubic Mn_3Si shows a complicated type of antiferromagnetism and is being discussed as an antiferromagnetic half-metal [128–130]. Mn_3Ge, which crystallizes in the hexagonal structure DO_{19}, exhibits a triangular spin configuration and weak ferromagnetism [131–134]. Similar magnetic properties were reported for Mn_3Sn [135, 136], which belongs to the same hexagonal structure type as Mn_3Ge. Another compound from this materials class is Mn_3Sb. This compound belongs to the cubic Cu_3Au structure type ($L1_2$) and shows weak ferromagnetism [137, 138].

Reducing the Mn content in such binary compounds, the compounds Mn_2Z are found. Mn_2Ge crystallizes in a hexagonal Ni_2In-type structure [139] and has not been investigated experimentally until now. Band structure calculations using the density functional theory led to the prediction of an antiferromagnetic ground state for Mn_2Ge [140]. Ferrimagnetic Mn_2Sb belongs to the tetragonal Cu_2Sb structure type, which is known as having a rich variety of magnetic structures [141, 142].

The present work focuses on the binary compound Mn_3Ga, which has long been an issue of interest because of its manifoldness regarding multiple structures and temperature-induced phase transitions. After some preliminary work on the binary Mn-Ga system [143, 144], Meißner and Schubert, in a comprehensive publication, established the framework for a phase diagram of this system [145]. Under suitable preparation conditions, three different phases are obtainable for the 3:1 composition of Mn and Ga. As-cast samples, obtained from repeated arc melting crystallize in a disordered Heusler-type DO_3 phase, which was predicted by Wurmehl *et al.* to show half-metallic completely compensated-ferrimagnetic behavior [146]. However, experiments provided evidence that the cubic phase of Mn_3Ga is not stable and cannot be produced without a high degree of disorder. Annealing of Mn_3Ga at high temperatures yields the hexagonal DO_{19} phase. A Mn-Ga composition with a Mn content of 71.25% was synthesized in the hexagonal phase in 1970 by Kren and Kadar [147]. This sample was investigated using neutron diffraction. Upon annealing this triangular antiferromagnetic structure at 750 K, a tetragonal DO_{22} phase was observed. This phase, which is the subject of this paper, is obtained by moderate annealing at a maximum temperature of 673 K. The phase is ferrimagnetic with magnetic moments of -2.8 μ_B for Mn_I and 1.6 μ_B for Mn_{II} (see Figure 10.1). Niida *et al.* [148] investigated several $Mn_{3-x}Ga$ compositions, but obtained single phase DO_{22}-type samples only in the range of $x = 0.15 - 1.16$. From *ab initio* electronic structure calculations, Kübler proposed a Curie temperature of $T_C = 762$ K for Mn_3Ga in the tetragonal DO_{22} structure [149]. The DO_{22} structure can be found by applying a tetragonal distortion to the DO_3 structure; therefore,

Mn$_3$Ga can be viewed as being a tetragonally distorted, binary Heusler compound and should exhibit properties that are associated with such a compound. Because of this, Balke *et al.* [17] recently suggested the applicability of Mn$_3$Ga in magnetoelectronic devices and, particularly, as a suitable material for spin torque transfer applications. The present work reports on the electronic, magnetic, and structural properties of Mn$_{3-x}$Ga with x varying from 0 to 1.0 in steps of $\Delta x = 0.1$. The electronic and magnetic structures were calculated by *ab initio* methods. The structure was investigated using powder X-ray diffraction (XRD), extended x-ray absorption fine structure (EXAFS) spectroscopy, and differential scanning calorimetry (DSC). The magnetic properties were studied using temperature and induction field dependent magnetometry. The transport properties of Mn$_3$Ga were investigated as to their dependence on temperature. The specific heats of Mn$_3$Ga and Mn$_2$Ga were investigated up to temperatures of 200 K and 150 K, respectively.

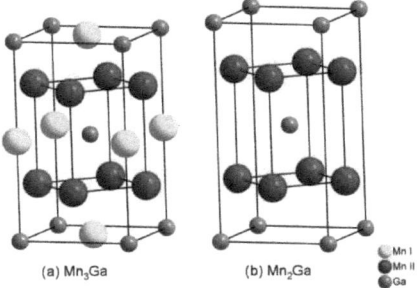

Figure 10.1: Structure of tetragonal Mn$_{3-x}$Ga. Mn$_3$Ga (from Ref. [17]) is shown in (a) and Mn$_2$Ga is shown in (b). The figure shows the DO_{22} crystal structure. Mn atoms (larger spheres) are represented by light (yellow) and dark (red) spheres, Ga atoms by gray spheres. In Mn$_3$Ga, the Mn$_I$ atoms are located on the base quadratic faces (1/2,1/2,0) and on the center plane (0,0,1/2); the Mn$_{II}$ atoms are found on the rectangular faces (0,1/2,1/4) and the Ga atoms at the corners and center of the cuboid. The Mn$_I$ atoms have a tetrahedral nearest neighbour environment. For Mn$_2$Ga shown in (b), only the Mn$_I$ atoms have been removed (see model I in Sec. IV A 2)); however, experiments indicate that both types of Mn atoms are removed (see text).

Calculational Details

Self-consistent electronic structure calculations of the electronic and magnetic properties of Mn$_3$Ga and Mn$_{3-x}$Ga were performed. The calculations were carried out using the spin polarized (SP) fully relativistic (R) Korringa-Kohn-Rostoker (KKR) method as implemented in the Munich-SPRKKR program [82, 150].

The exchange-correlation functional was taken within the parametrization of Vosko, Wilk, and Nussair [83, 84]. For comparison, the generalized gradient approximation (GGA) in the parametrization of Perdew et al. [75] was used; however, no remarkable differences were found between either parameterization. A base, $14 \times 14 \times 17$ mesh, was used for integration. The mesh corresponds to 368 (out of 3332) k points in the irreducible wedge of the Brillouin zone. The size of the muffin-tin radii was set automatically to result in space-filling spheres, and depends on the lattice parameters. f states ($l = 3$) were included in the basis of both atoms, Mn and Ga.

The properties of $Mn_{3-x}Ga$ were calculated in $I\,4/mmm$ symmetry using polynomial fits of a and c to the experimental lattice parameters (see Section 10). In the DO_{22} structure type, the Mn atoms occupy two different positions. The first position (Mn_I), with multiplicity 1, is located at the Wyckoff position 2b (0,0,1/2) and the second position (Mn_{II}), with multiplicity 2, is at 4d (0,1/2,1/4). The Ga atom is at the Wyckoff position 2a (0,0,0). The lack of manganese in $Mn_{3-x}Ga$ was simulated by introducing vacancies (◇) in the DO_{22} structure. To account for the random distribution of the vacancies for $x \neq 0$, the calculations were carried out using the coherent potential approximation (CPA).

No essential changes of the electronic and magnetic structures of Mn_3Ga were obtained when the calculations were performed for the optimized lattice parameter as reported in Ref. [17].

Experimental Details

$Mn_{3-x}Ga$ samples were prepared by repeated arc melting of stoichiometric amounts of the constituents in an argon atmosphere at 10^{-4} mbar. Care was taken to avoid oxygen contamination. The resulting polycrystalline ingots were annealed afterward at 623 K in an evacuated quartz tube for 14 days. The temperature was chosen to be safely below the structural transition temperature (see Section 10) and to be high enough to ensure sufficient diffusion. After annealing, the samples were quenched in an ice/water mixture so that the desired structure would be retained. This procedure resulted in samples exhibiting the DO_{22} structure (space group: $I\,4/mmm$) as was confirmed by using powder X-ray diffraction. The unit cell of the DO_{22} structure is displayed in Figure 10.1.

DSC measurements (NETZSCH, STA 429) were performed to detect phase transitions below the melting points of the materials. The crystal structure of the $Mn_{3-x}Ga$ series was determined using XRD with excitation by Mo K_α radiation. In addition, temperature dependent XRD was used for a detailed structural investigation of Mn_3Ga.

The latter powder XRD experiments were carried out at the D10B-XPD beamline of the LNLS (Brazilian Synchrotron Light Laboratory, Campinas, Brazil).

The short range order of Mn$_3$Ga was investigated using EXAFS. The EXAFS experiments were performed at the XAFS1 beamline [151] of the LNLS using a Si(111) channel-cut monochromator. The spectra were collected at room temperature in the transmission mode at the Mn (6539 eV) and Ga (10367 eV) K edges using three ionization chambers. Ga or Mn standard foils were placed at the third chamber to check the monochromator energy calibration. The EXAFS spectra were analyzed using the IFEFFIT analysis package [152]. The isolated atom background function was removed from the experimental X-ray absorption coefficient data to yield the $\chi(k)$ signal. The Fourier transform (FT) was applied using a Hanning window with a k range of 8 Å^{-1}. The structural parameters were obtained from a least-squares fit to the data in r and k space using phase shifts and amplitudes that were obtained from the FEFF code [153] as calculated for Mn$_3$Ga in the DO_{22} structure and accounting for the two possible Mn sites shown in Figure 10.1.

The magnetic properties were investigated using a superconducting quantum interference device (SQUID, Quantum Design MPMS-XL-5) using small spherical sample pieces of approximately 5-10 mg of the sample. The transport properties and the specific heats were investigated using a physical property measurement system (Quantum Design PPMS models 6000 and 6500, respectively).

Results and Discussion

Calculated electronic structure

Mn$_3$Ga.

For the parent compound Mn$_3$Ga, the *ab initio* calculations indicated a ground state with ferrimagnetic order for the fully occupied DO_{22} cell. The spin magnetic moments (m_s) found from the calculations using the VWN parametrization are -3.049 μ_B for Mn$_I$ and 2.386 μ_B for Mn$_{II}$. The orbital moments are parallel to the spin moments and amount to -0.02 μ_B and 0.03 μ_B for Mn$_I$ and Mn$_{II}$, respectively. The total magnetic moment of 1.702 μ_B in the primitive cell is low due to the partial compensation of the moments of the Mn atoms on the two different sites. Using the GGA approach, the total magnetic moment is found to be slightly higher (1.83 μ_B). A particular enhancement of the orbital moments is not observed; they amount to approximately 1% of the spin moments, independent of the exchange-correlation functional used. The values for total and local moments agree well with those previously reported for full potential calculations in Ref. [17].

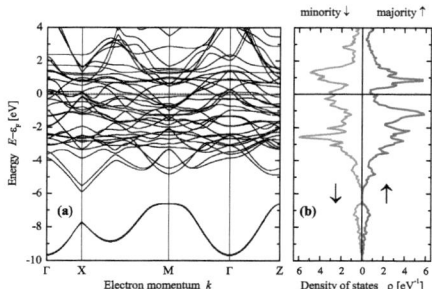

Figure 10.2: Electronic structure of Mn$_3$Ga. The relativistic band structure is shown in (a), along with the spin resolved density of states in (b).

Figure 10.2 shows the full relativistic band structure and the spin resolved density of states (DOS) for Mn$_3$Ga. Unlike the nonrelativistic approach (see Ref. [17]), the minority and majority character of the band structure cannot be distinguished. This is only possible in the density of states. The reason is the intrinsic coupling of the spin states because of the spin-orbit interaction in the Dirac equation. The density of states, as an expectation value, can still be determined with respect to the direction of the spin. As was already discussed from the results of nonrelativistic calculations, the higher density of minority states at the Fermi energy ϵ_F (compared to the low majority density) suggests that there is a remarkable difference in the conductivity between the minority and the majority electrons as was already discussed from the result of nonrelativistic calculations [17]. Typical for a large family of compounds between transition metals and main group elements is the occurrence of a hybridization gap that splits the low lying s bands (at below approximately -7 eV) from the high lying p and d bands. The size of that gap, appearing in both spin directions, decreases with decreasing hybridization strength. This is similar to what is observed for the related Heusler compounds (see Ref. [62]).

Mn$_{3-x}$Ga, $0 < x \leq 1.0$.

Basically, two different structural models were employed in the calculations for the compounds with a lack of Mn. In the first model, only the Mn$_I$ was removed from the *2b* position resulting in Mn$_2$(Mn$_{1-x}\diamond_x$)Ga (model I). In the second model, the Mn was removed from both positions simultaneously (*4c* and *2b*), resulting in (Mn$_{2-2x/3}\diamond_{2x/3}$)(Mn$_{1-x/3}\diamond_{x/3}$)Ga (model II). The first model is based on the experimental observation (see Section 10) that the magnetic moment increases as the lack of Mn increases, and is higher in Mn$_2$Ga than in Mn_3Ga. In the ferrimagnetic state the moments of Mn_I and Mn_{II} are aligned in antiparallel fashion. Therefore, at least for fixed local moments, the removal of Mn_I will increase the total magnetic moment.

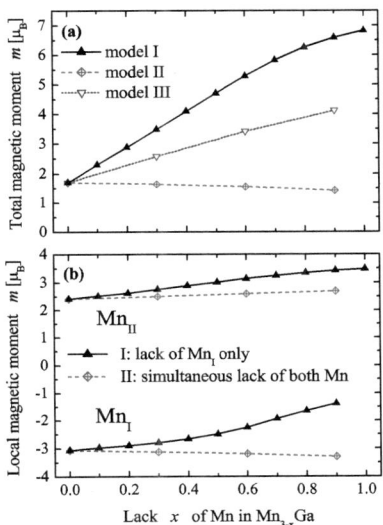

Figure 10.3: Calculated magnetic moments of $Mn_{3-x}Ga$. The total magnetic moment in the DO_{22} cell is shown in (a) and the local magnetic moments at the Mn_I and Mn_{II} sites are given in (b). Note that the magnetic moment $m = m_l + m_s$ contains both spin (m_s) and orbital (m_l) moments. See text for an explanation of the different models (I - III) used in the calculations.

The calculated magnetic moments for the two structural models are compared in Figure 10.3. In contrast to experimental observation, the total moment decreases slightly with increasing lack x of Mn for model II [$(Mn_{2-2x/3}\diamond_{2x/3})(Mn_{1-x/3}\diamond_{x/3})Ga$]. The reason is clear. Although the local moments do not stay constant, their influence is partly compensated in the total moment that is mainly given by the composition $m = (1 - x/3)\{2m_{II} - |m_I|\}$. The behavior of the total magnetic moment in model I [$Mn_2(Mn_{1-x}\diamond_x)Ga$] follows the expected path; it increases with increasing lack of manganese. The simultaneous increase of the moment at Mn_{II} and the decrease of the absolute value of the moment at Mn_I cause a strong increase in the overall moment.

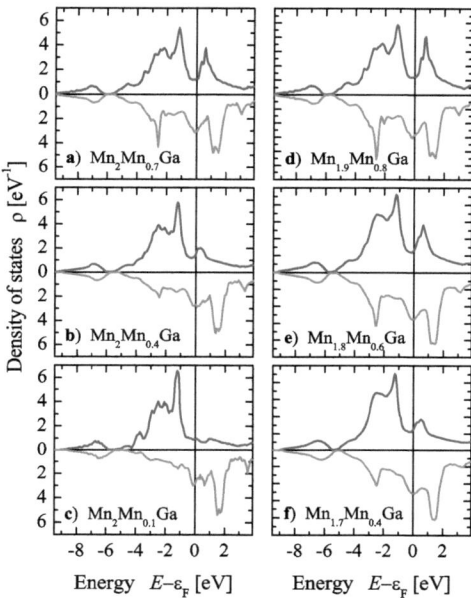

Figure 10.4: Density of states for $Mn_{3-x}Ga$. (a)-(c) show the DOS for model I: $Mn_2(Mn_{1-x}\diamond_x)Ga$ (d)-(f) show the DOS for model III: $Mn_{2-x/3}\diamond_{x/3})(Mn_{1-2x/3}\diamond_{2x/3})Ga$. Shown are the selected compositions with $x = 0.3$ [(a) and (d)], 0.6 [(b) and (e)], and 0.9 [(c) and (f)]. Majority densities are shown in the upper parts, and minority states in the lower parts of each of the plots.

However, the increase with x of the total magnetic moment of model I is more pronounced than that found in the experimental data. Therefore, a third model was introduced where more Mn_I is removed with increasing x than Mn_{II}. The composition in model III was chosen as $(Mn_{2-x/3}\diamond_{x/3})(Mn_{1-2x/3}\diamond_{2x/3})Ga$, which is an intermediate case between the two other models, which constitute the borderline cases of Mn removal from the crystallographic sites. For this model, twice as much Mn_I is removed as Mn_{II}. From Figure 10.3(a), it is apparent that the increase of the total magnetic moment is weaker than for the $Mn_2(Mn_{1-x}\diamond_x)Ga$ model. Further details will be discussed later, together with the experimental data.

Figure 10.4 shows the DOS for the two structural models I and III, where mainly

Mn_I is removed with increasing x. To cover most of the cases, the compositions with $x = 0.3$, 0.6, and 0.9 were chosen for the comparisons. The VWN parametrization of the exchange-correlation functional was used in the calculations.

The general shape of the density of states is quite similar for all the cases. In particular, the minority DOS exhibits a maximum at the Fermi energy. At the same time, the majority DOS has a minimum at ϵ_F that is less pronounced at low Mn content and vanishes in the particular case of $Mn_2Mn_{0.1}Ga$. The latter result is typical for model I, where the maximum of the DOS above ϵ_F is decidedly lowered with the removal of the Mn_I atoms. At the same time the high density of states in the minority channel at approximately -3 eV is lowered as Mn_I atoms are removed. This explains why the localized moment of the Mn_I atom is pronouncedly reduced upon removal of those atoms from the structure. The more delocalized moment of the Mn_{II} atom is less affected. In all cases, a decrease in the size of the low lying band gap at about -6 eV that splits off the s-p states from the d states is observed. This indicates a weaker hybridization between Mn and Ga in the compounds with low Mn content.

Structural characterization

X-ray powder diffraction

Table 10.1: Lattice parameter of $Mn_{3-x}Ga$. As found from a Rietveld refinement of the XRD data, the tetragonal lattice parameters a and c are tabulated for decreasing Mn content $(3-x)$. The ratio c/a and volume V are calculated from the measured values of a and c. The maximum errors for the lattice parameters are estimated from the (200) and (004) reflection peaks to be $\Delta a = 0.004$ Å and $\Delta c = 0.016$ Å. (Uncertainty for calculated quantities follows from error propagation.)

x	a [Å]	c [Å]	c/a	V [Å^3]
0	3.909	7.098	1.816	54.214
0.1	3.907	7.100	1.817	54.118
0.2	3.908	7.105	1.818	54.265
0.3	3.910	7.117	1.820	54.396
0.4	3.910	7.122	1.821	54.489
0.5	3.909	7.130	1.824	54.467
0.6	3.906	7.155	1.832	54.581
0.7	3.905	7.160	1.834	54.602
0.8	3.906	7.169	1.835	54.680
0.9	3.906	7.175	1.837	54.719
1	3.905	7.193	1.849	54.930

The crystalline structure of the $Mn_{3-x}Ga$ series was examined using XRD with excitation by Mo K_α radiation. The measurements were performed at room temperature. As examples, Figure 10.5 shows the XRD results for three selected compositions ($x = 0.1, 0.5$ and 1.0). The XRD data verify that, for all the compositions, the $Mn_{3-x}Ga$ samples crystallize in the Al_3Ti structure type, which belongs to the space group $I 4/mmm$. The diffraction patterns for all the compositions (except for Mn_2Ga) did not show any unexpected reflexes. This indicates the phase purity of the samples. However, a small shoulder on the low angle side of the (112) reflex in the pattern for Mn_2Ga indicates an impurity, which could not be dedicated.

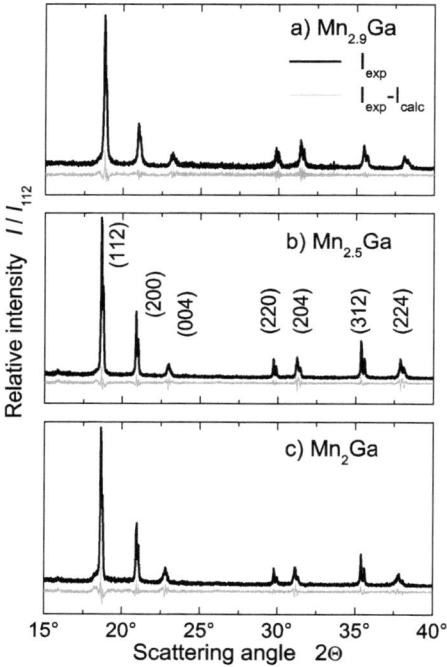

Figure 10.5: Powder diffraction of $Mn_{3-x}Ga$. The measured diffractograms for $Mn_{2.9}Ga$, $Mn_{2.5}Ga$, and Mn_2Ga are compared to the curve fits from a Rietveld refinement. The spectra were excited using Mo K_α radiation and recorded at $T = 300$ K.

The lattice parameters found from a Rietveld refinement of the powder XRD data are summarized in Table 10.1. Within the uncertainty of the measurements, the a parameter remains constant when the Mn content in the samples is reduced. At the same time, the c parameter increases by approximately 1.7% as x varies from 0 to 1.0. Overall, this results in an increase of the cell volume V as well as the c/a ratio with decreasing Mn content. A removal of Mn$_I$ atoms will apparently weaken the Mn-Ga bonds along the c axis. The increase of the c-parameter with x also indicates that more Mn$_I$ atoms than Mn$_{II}$ atoms are removed along the c axis when the Mn content is reduced. The Rietveld refinement is compatible with both of the models used for the calculations: (I) for the removal of Mn$_I$ only, as well as (III) for the removal of a higher amount of Mn$_I$ than Mn$_{II}$.

Temperature dependent XRD

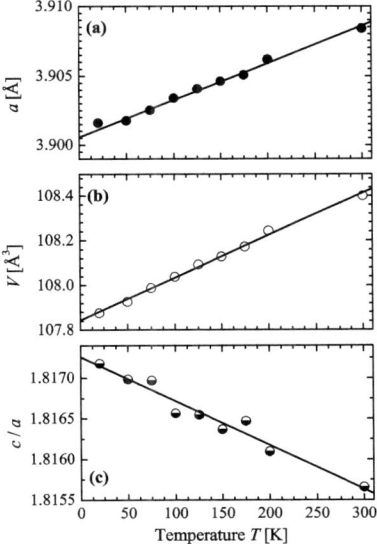

Figure 10.6: Temperature dependence of the Mn$_3$Ga lattice parameters. The temperature dependence of the lattice parameter a and the cell Volume V is shown in (a) and (b), respectively. The temperature dependence of the ratio c/a of the tetragonal lattice parameters is shown in (c).

The crystal structure was recorded in the temperature range between 25 K and 200 K using a step size of 25 K and, additionally, at 300 K using excitation by radiation with a wavelength of 1.75866 Å. Figure 10.6 shows the temperature dependence of the lattice parameter a, the volume of the tetragonal unit cell V and the c/a ratio. It is seen that the lattice parameter a and the cell volume decrease linearly with decreasing temperature. The lattice parameter c, which is not shown in Figure 10.6, also exhibits a linear decrease when the temperature is reduced. However, Figure 10.6(c) indicates that the c/a ratio increases when the temperature decreases. When the temperature is raised, the lattice parameter a, thus, grows stronger than the lattice parameter c. The decrease of c/a with T means that the tetragonal unit cell of Mn_3Ga becomes closer to a cubic structure as the temperature increases. However, the ratio c/a, even at room temperature, is far from the ideal value ($\sqrt{2}$) that is expected for a cubic system with the DO_3 structure.

Extended x-ray absorption fine structure

The results from the analysis of the EXAFS measurements are summarized in Figure 10.7. The EXAFS signals [$\chi(k)$] at the Mn and Ga [Figure 10.7(a)] K edges display the characteristic patterns for a cubic structure. The fit of the FT curves for both elements is given in Figure 10.7(b). Excellent agreement between the data and theory is found as demonstrated by the low R-factor values (between 0.01 and 0.02). The Fourier transform at the Ga K-edge spectra exhibits one peak at approximately 2.2 Å and another at 4.5 Å (not corrected for the scattering phase shift) that correspond mainly to the Ga-Mn single scattering contribution in the coordination shell and the Ga-Ga contribution, respectively. Other scattering contributions appear in the quite smooth region between the two main peaks.

The FT at the Mn K-edge spectra exhibits peaks at similar positions but with lower amplitudes. In fact, the Mn atoms have two possible sites: site Mn_I surrounded by 8 Mn atoms, and site Mn_{II} surrounded by (4 Mn + 4 Ga) atoms. The peak at about 4.5 Å corresponds mainly to the Mn-Ga and Mn-Mn contributions. Because the coordination number for Ga and Mn are the same, the difference in intensity may be explained by disorder effects.

The curve fitting procedures for Mn_3Ga lead to physically reasonable numbers that are close to those provided by the theoretical structural model. The obtained values for the passive electron reduction factor (S_0^2) are about 0.7. The shift in distance (ΔR) and the Debye-Waller factor (σ^2) with respect to the theoretical model for the paths involving Ga-Ga or Ga-Mn_I(Mn_{II}) were small throughout, with $\Delta R \approx 0.04$ Å and $\sigma^2 \approx 0.007$ Å^2. For the paths involving only Mn atoms, Mn_I(Mn_{II}) - Mn_I(Mn_{II}), the ΔR and σ^2 values were considerably higher, with $\Delta R \approx 0.13$ Å and $\sigma^2 \approx 0.02$ Å^2, respectively. This is in agreement with the lower amplitudes obtained at the Mn K edge. The curve fitting results clearly indicate the DO_{22} structure for the Mn_3Ga composition.

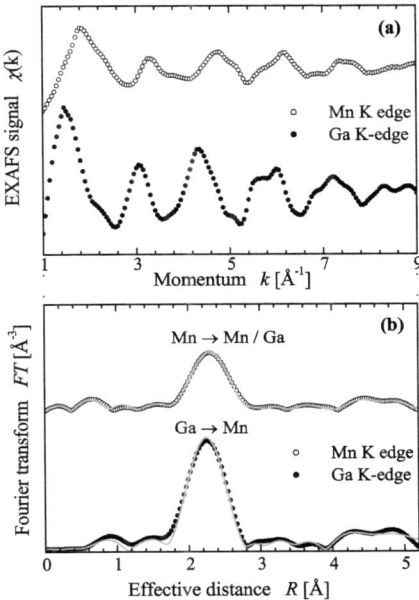

Figure 10.7: EXAFS results for Mn_3Ga. EXAFS oscillations at the Mn and Ga K edges, extracted from the x-ray absorption spectra, are given in (a). The corresponding Fourier transforms (symbols) and best fit results (gray line) are given in (b).

Structural phase transitions

Using XRD, it was observed that the samples undergo a transition to the hexagonal DO_{19} structure at high temperatures. For this reason, DSC measurements of the entire sample series were performed to search for and investigate possible phase transitions. The DSC signals were recorded in the range from 300 K to 1250 K for heating and cooling rates of 5 K min^{-1} to examine the reversibility of the occurring phase transitions. As an example, Figure 10.8(a) shows two DSC curves for the compositions $Mn_{2.3}Ga$ and $Mn_{2.7}Ga$.

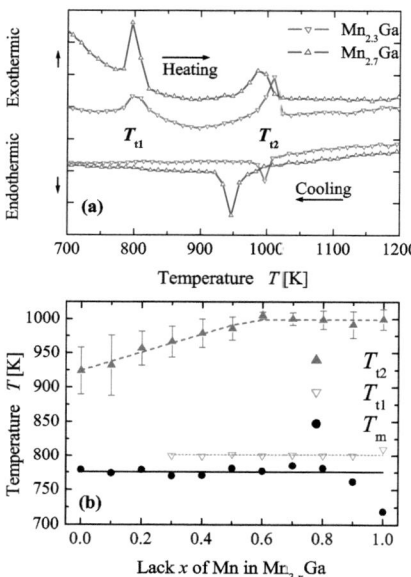

Figure 10.8: Phase transitions in $Mn_{3-x}Ga$. The plots in (a) show the results from differential scanning calorimetry for $Mn_{2.7}Ga$ and $Mn_{2.3}Ga$ in the temperature range between 700 K and 1200 K upon heating of the samples, followed by cooling. The composition dependence of the phase transition temperatures is summarized in (b). T_{t1} and T_{t2} are taken from the DSC measurements, T_m is from high temperature magnetization measurements (see Section 10). (Lines are drawn for clarity.)

The change of the DSC signal as a function of temperature is given. Upon heating, both substances exhibit a first maximum at a temperature T_{t1} of approximately 800 K that is not observed upon cooling, thus indicating an irreversible phase transition. In addition, a second, reversible phase transition is observed at higher temperatures, which is assigned a temperature of T_{t2}. It is seen from Figure 10.8(b) that the phase transition temperature T_{t1} is independent of composition. Furthermore, it is evident that the signal at about 800 K becomes less intense and is broadened with increasing Mn content, a trend that was observed throughout the entire series. For $x \leq 0.2$, the determination of this phase transition at T_{t1} was impossible because the levels of the signals were too low. Nevertheless, the magnetic measurements (Section 10) also indicate that there is a magnetic phase transition between 750 K and 830 K in all the samples. The phase transition observed using DSC at T_{t1} corresponds to a transition to the high temperature hexagonal phase DO_{19} (structure type: Ni_3Sn, space group $P\,6_3/mmc$). The structural nature of the phase transition was verified experimentally by annealing the samples at a temperature of 873 K and then examining the crystal structure using XRD. From Figure 10.8(b), it is seen that the signals at temperature T_{t2} exhibit hysteresis between heating and cooling that is due to the intrinsic effects of DSC. The shift between the extrema depends on the amount of material used as well as on the rates of temperature change. Therefore, the mean value of T_{t2} is displayed in Figure 10.8(b) with error bars indicating the differences between the endothermic and exothermic values. The phase transition appears throughout the entire range of sample composition. It is clear that T_{t2} increases with decreasing Mn content and saturates for $x \geq 0.6$ at 1000 K. It is very likely that this corresponds to an order$\leftrightarrow$disorder transition in the hexagonal phase with preserved symmetry. Such a transition would hardly be detectable with XRD because the scattering amplitudes of Mn and Ga have similar magnitudes. It appears that this high temperature phase transition at T_{t2} changes the sample properties in such a way that the first phase transition at approximately 800 K can no longer occur upon cooling.

Magnetic properties

The results of the induction field dependent magnetization measurements are displayed in Figure 10.9. All samples exhibit hard magnetic behavior at temperatures below the phase transitions to the hexagonal DO_{19} phase. For $Mn_{2.9}Ga$, hysteresis loops at three different temperatures are displayed in Figure 10.9(a). It is seen that the magnetic moment is stable up to a temperature of 400 K. For the compounds Mn_3Ga and $Mn_{2.9}Ga$, Table 10.2 summarizes the values of the coercive fields H_c, remanences B_r, the specific energy products $E = H_c \times B_r$ and the energy integrals $W_H = \oint H dB$ at temperatures of 5 K, 300 K, and 400 K, respectively. The energy products E of Mn_3Ga and $Mn_{2.9}Ga$ are comparable to those of other hard magnetic materials such as ALNICO 6. But the coercitivities of Mn_3Ga and $Mn_{2.9}Ga$ are approximately four to five times higher than that of ALNICO 6. The relatively low saturation magnetizations, due to the ferrimagnetic order, make the materials comparable to ferrites and related materials. From a direct integration of the magnetization loops, the energy integral $W_H = \oint H dB$ gives the hysteresis loss per cycle. For Mn_3Ga and $Mn_{2.9}Ga$, it is between three and four times larger than the energy product. For an ideal hard magnet, a nearly rectangular hysteresis loop $(B(H))$ with $W_H \approx 4E$ and $E = BH_{max}$ is expected. The maximum energy product $BH_{max} = \max(-B \times H)$ is an energy parameter of interest for the manufacture of magnets. It represents the maximum useful magnetic energy of a permanent magnet. Its value is obtained by multiplying B times H in the second (or fourth) quadrant of the hysteresis loop. Figure 10.10 shows a plot of BH and the maximum energy product for the case of Mn$_3$Ga. The value of $BH_{max} \approx 18$ kJm^{-3}, like the other energy values, is comparable to the value found in magnets from the ALNICO series.

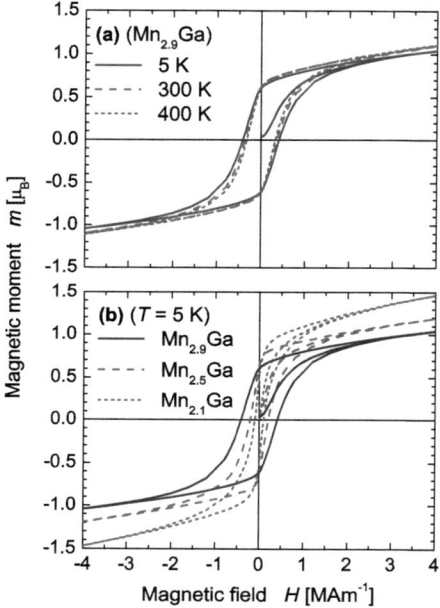

Figure 10.9: Magnetic properties of Mn$_{3-x}$Ga. The field dependent magnetization for Mn$_{2.9}$Ga at temperatures of 5 K, 300 K, and 400 K is given in (a). Hysteresis loops for Mn$_{2.9}$Ga, Mn$_{2.5}$Ga and Mn$_{2.1}$Ga at $T = 5$ K are given in (b).

The hysteresis loops for Mn$_{2.9}$Ga, Mn$_{2.5}$Ga and Mn$_{2.1}$Ga are given in Figure 10.9(b). The curves reveal a trend that is observed throughout the entire sample series. The coercivity H_c decreases upon removal of Mn (increasing x). The remanence B_r does not strongly dependent on x. Since complete saturation is not reached up to an induction field of 5 T for all samples, the value of the saturation magnetization cannot be determined. However, the maxima of the magnetic moments (at 5 T) of the series clearly increase with increasing x. This accords well with the structural properties

10. Structural, electronic, and magnetic properties of tetragonal $Mn_{3-x}Ga$: Experiments and first-principles calculations

(Table 10.1). The increase of the lattice parameter c with increasing x, which is caused by the removal of the Mn_I atoms along the c-axis, leads to a reduction of the magnetic compensation in the ferrimagnetic unit cell of Mn_3Ga. This explains why, at a given induction field, the magnetic moments increase when Mn is removed from Mn_3Ga.

Table 10.2: Hard magnetic properties of Mn_3Ga and $Mn_{2.9}Ga$. Values are provided at low and elevated temperatures T of the coercive field H_c, the remanence B_r, the specific energy product E, the energy integral W_H and the maximum energy product BH_{max}. (Errors for the measured quantities are: $\Delta H_c = 2$ kAm^{-1} and $\Delta B_r = 0.5$ mT. Errors for the calculated quantities E, W_H and BH_{max} may be found using error propagation.)

	T [K]	B_r [T]	H_c [kAm^{-1}]	E [kJm^{-3}]	W_H [kJm^{-3}]	BH_{max} [kJm^{-3}]
Mn_3Ga	5	0.136	453	61.6	203	18.3
	300	0.133	383	50.9	176	14.4
	400	0.130	342	44.5	149	12.7
$Mn_{2.9}Ga$	5	0.131	412	54.0	171	15.4
	300	0.132	359	47.4	158	13.9
	400	0.131	322	42.2	140	12.2

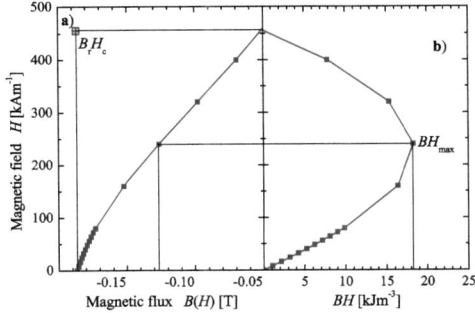

Figure 10.10: BH_{max} of Mn_3Ga. The second quadrant of the hysteresis loop $B(H)$ recorded at $T = 5$ K is shown in (a). The calculated values for $BH(H) = -B(H) \times H$ are given in (b). The maximum value BH_{max} of approximately 18 kJm^{-3} is found for a magnetic field of ≈ 250 kAm^{-1} and a flux density of ≈ 0.07 T. [Note that the x (H) and y (B, BH) axes are interchanged for better comparison.]

Calculated and measured magnetization data are compared in Figure 10.11. When going from Mn_3Ga to $Mn_{2.1}Ga$, an increase in magnetization of about 80% is found experimentally. The increase in the calculated values is much larger ($\approx 300\%$) for model I where only Mn_I is removed. According to model III, the simultaneous removal of Mn_I and Mn_{II} results in a reduced increase of 140% if it is assumed that twice as much Mn_I as Mn_{II} is removed. Table 10.3 provides an overview of the physical properties of the $Mn_{3-x}Ga$ series. It is evident that varying the Mn content allows a simultaneous continuous tunability of coercivity, saturation magnetization and energy product, making the desired properties of the materials easily adjustable.

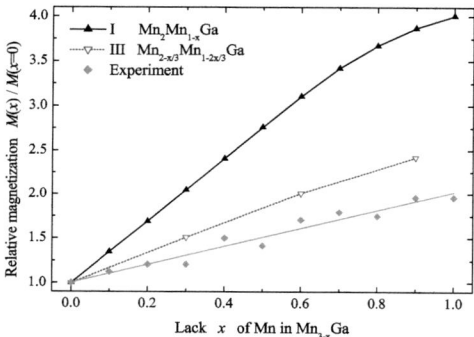

Figure 10.11: Comparison of magnetization data for $Mn_{3-x}Ga$. The values are normalized to $x = 0$ for better comparison. Experimental values are for an induction field of 5 T. Calculated values for models I and III are seen in Section 10.

The (mass) density ρ of $Mn_{3-x}Ga$ decreases drastically by about 30% with increasing x. This is because the number of Mn atoms in the unit cell decreases while the volume of the unit cell increases with x. The decrease is, nevertheless, linear and a fit results in $\rho(x) = (1 - 0.2441\,x) \times 7189$ kgm^{-3}. This means that the decrease of the density is governed by the loss of mass since a slope of about -0.2342 is expected for a fixed volume. The magnetic transition temperature is nearly constant with a mean value of $\overline{T_m} = (775 \pm 2)$ K (omitting the value at $x = 1$). The comparatively low transition temperature for Mn_2Ga may be attributed to the impurity that was found from XRD measurements (see Section 10). From Table 10.3, it is apparent that $Mn_{3-x}Ga$ becomes magnetically softer with decreasing Mn content. All three energy parameters E, W_H and BH_{max} decrease with increasing x. At the same time, a decrease of the coercivity and a maximum of the remanence at intermediate Mn concentrations are observed.

10. Structural, electronic, and magnetic properties of tetragonal $Mn_{3-x}Ga$: Experiments and first-principles calculations

Table 10.3: Magnetic properties of $Mn_{3-x}Ga$. The following are tabulated: density ρ (at $T = 300$ K), magnetic moment m per atom (at $\mu_0 H = 5$ T), the high temperature magnetic transition T_m, remanence B_r, coercive field H_c, specific energy product E, energy integral W_H and the maximum energy product BH_{max}. The magnetic properties are given for 5 K. (Maximum errors were determined to be: $\Delta m = 5 \times 10^{-3} \mu_B$, $\Delta T_m = 5$ K, $\Delta B_r = 0.5$ mT and $\Delta H_c = 2$ kAm^{-1}. Other uncertainties are found from error propagation.

x	ρ [kgm^{-3}]	m [μ_B]	T_m [K]	B_r [T]	H_c [kAm^{-1}]	E [kJm^{-3}]	W_H [kJm^{-3}]	BH_{max} [kJm^{-3}]
0	7184	0.26	779	0.137	453	61.6	203	18.3
0.1	7028	0.27	774	0.131	412	54.0	171	15.4
0.2	6841	0.29	779	0.137	344	47.1	159	12.3
0.3	6657	0.29	770	0.130	315	41.0	146	10.4
0.4	6478	0.36	771	0.155	249	38.6	150	8.8
0.5	6313	0.34	781	0.139	205	28.5	122	6.8
0.6	6133	0.41	777	0.127	149	18.9	116	3.8
0.7	5963	0.43	785	0.125	137	17.1	114	3.4
0.8	5788	0.42	781	0.114	124	14.1	105	2.9
0.9	5617	0.47	762	0.111	100	11.1	91	2.6
1	5429	0.47	723	0.099	102	10.1	90	2.4

The temperature dependence of the magnetization is shown in Figure 10.12 for the cases of Mn_3Ga and Mn_2Ga. In the low temperature region, the measurements were performed using induction fields of 0.1 T under zero-field-cooled (ZFC) and field-cooled (FC) conditions. For the ZFC measurements the samples were first cooled to a temperature of 1.8 K without applying a magnetic field. After applying an induction field $\mu_0 H$, the sample magnetization was recorded as the samples were heated to the maximum temperature. Immediately afterwards, the FC measurements were performed in the same field as the temperature was lowered to 1.8 K. The results for low temperatures are shown in Figures 10.12(a) and (b). Evidently, the curves show no regular ferro-, antiferro- or ferrimagnetic behavior. Magnetic transitions are observed at temperatures of 164 K for Mn_3Ga and approximately 145 K for Mn_2Ga in the ZFC measurements. These transitions exhibit thermal irreversibilities between the ZFC and the FC measurements. This type of behavior is known from spin glasses and other frustrated systems with multi-degenerate ground states [39]. Thermal irreversibilities, such as the ones mentioned above, are found in all samples of the entire series. Furthermore, the irreversibilities were observed using different magnetic induction fields that ranged from 2.5 mT to 5 T. These irreversibilities lead to the conclusion that the $Mn_{3-x}Ga$ series is a set of frustrated ferrimagnets [24, 154].

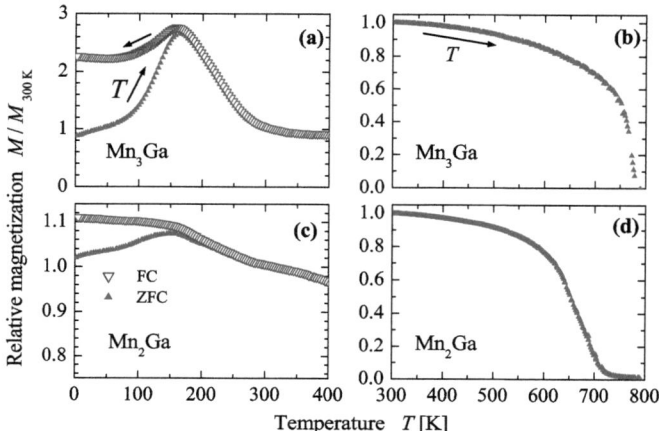

Figure 10.12: Temperature dependent magnetization of Mn_3Ga and Mn_2Ga. The temperature measurements in the range from 5 K to 400 K under ZFC and FC conditions are given in (a) and (b). The high temperature measurements ($T > 300$ K) upon heating of the samples are given in (c) (from Ref. [17]) and (d). For better comparison, the data are normalized to the values at 300 K [note the offset of the Mn_2Ga low temperature magnetization data in (c)].

The difference of the magnetization between ZFC and FC measurements decreases with increasing x. For Mn_3Ga, there is a difference of 131% between ZFC and FC measurements while the same difference constitutes only 9% for the case of Mn_2Ga. Apparently, the magnetic frustration is lowered upon removal of Mn atoms from the unit cell. This also supports model III. According to model I, where only Mn_I atoms are removed in Mn_2Ga, a ferromagnetic order should remain. The remaining magnetic frustration in Mn_2Ga thus provides evidence that Mn_2Ga contains both Mn_I and Mn_{II} atoms with an at least partially random distribution among both crystallographic sites. For some intermediate compositions, an accurate determination of the transition temperatures was not clearly possible because the transitions were very broad or there was no distinct maximum at all (in the cases of $x = 0.2$ and 0.4). The transitions that could be determined span a temperature range from 145 K to 172 K. It appears, however, that there is no distinguished correlation of the transition temperatures with x, the magnetic moment or the lattice parameters. This hints also on a partial random distribution of the voids on both Mn positions.

The drastic decreases of the magnetizations at $T_S \approx 780$ K for Mn_3Ga and at $T_S \approx 720$ K for Mn_2Ga, which are seen in the high temperature SQUID measurements in Figure 10.12(c) and (d), are caused by structural transitions from the tetragonal to the hexagonal phases [145, 147]. These structural transitions are easily detected using differential scanning calorimetry. The occurrence of the structural phase transitions makes it impossible to determine the Curie temperature of the tetragonal phase of $Mn_{3-x}Ga$. The structural phase transitions take place at temperatures below the expected magnetic phase transition. The high temperature magnetic measurements also support the results from DSC (see Section 10). The lower temperatures of T_m compared to T_{t1} [see Figure 10.8(b)] are addressed to an intrinsic effect of the DSC, where the signals depend on the heating rates; higher heating rates and larger amounts of material will apparently shift the signal to higher temperatures. The magnetic measurements also show the onset of magnetization at T_m upon cooling. This confirms the assumption that T_m and T_{t1} correspond to the magnetic transition temperature of the hexagonal DO_{19} phase rather than to the Curie temperature of the tetragonal DO_{22} phase. The magnetic transition and the structural transition occur nearly simultaneous. Nevertheless, this should not cause problems in designing devices using this material because it is very stable in the temperature range below 700 K and, thus, well suited for applications at and far above room temperature.

Transport properties

The resistance of Mn_3Ga and Mn_2Ga was measured as a function of temperature upon heating and cooling in the range from 2 K to 300 K. The heating curves are shown in Figure 10.13. The values are normalized to the value at 2 K for better comparison. The samples show typical metallic behavior, where the resistances increase with increasing temperature. At approximately room temperature, the resistances of Mn_3Ga and Mn_2Ga exhibit a linear slope. Only a very small thermal hysteresis was found between the heating and cooling curves and it can be attributed to the precision of the measurements. Residual resistivity ratios of approximately 8 for Mn_3Ga and 4.5 for Mn_2Ga were determined for the ratio between the 2 K and 300 K measurements. For polycrystalline ingots, the comparably large value of 8 indicates the high quality of the Mn_3Ga samples. However, the value for Mn_2Ga is significantly lower. This fact is addressed to the unknown impurity, which was found in the XRD measurement (Section 10). The low temperature magnetic transitions, as mentioned in Section 10, are not observable in the resistivity measurements. It is known from other frustrated systems that the transitions may be shifted to much higher temperatures (values of 150% are quite common, shifts of up to more than 1000% are known) or are simply not detectable in the resistivity measurements [39, 155–157]. The latter is the most probable reason because a sharp phase transition is also not detectable in the magnetic measurements.

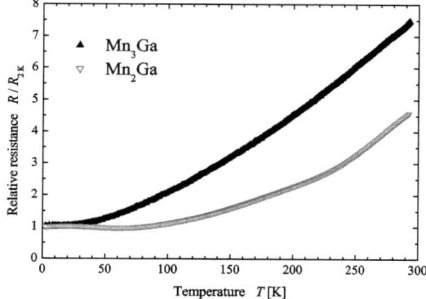

Figure 10.13: Electric resistance measurements of Mn_3Ga and Mn_2Ga. The measurements were performed upon heating of the samples. For better comparison, the data are normalized to the values at 2 K.

Specific heat

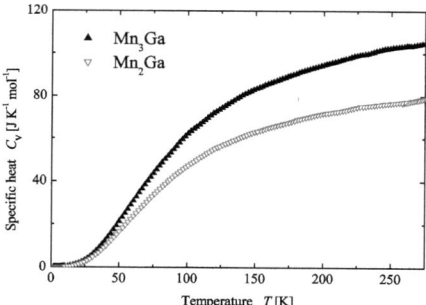

Figure 10.14: Specific heat of Mn_3Ga and Mn_2Ga. The measurements were performed upon heating of the samples.

The specific heat of Mn_3Ga and Mn_2Ga was measured as a function of temperature. The results are displayed in Figure 10.14. The curves show typical metalliclike specific heat shapes composed of electronic and phononic contributions to the specific heats. To a good approximation, they follow T^3 laws below 30 K. A low temperature fit (< 5 K) of the data results in electron specific heats of 19 ± 0.2 mJ/(molK^2) and 16 ± 0.2 mJ/(molK^2) for Mn$_3$Ga and Mn$_2$Ga, respectively. Those values are rather large compared to pure Mn [$\gamma_{Mn} = 9.2$ mJ/(molK^2)].

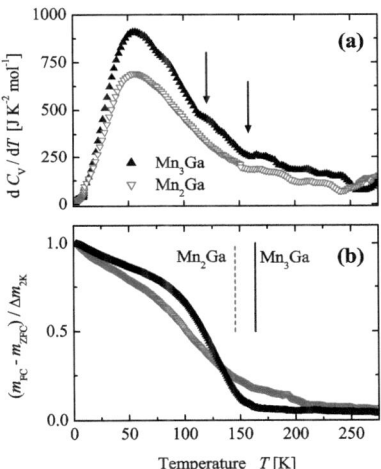

Figure 10.15: Low temperature magnetic anomaly of Mn$_3$Ga and Mn$_2$Ga. (a) shows the derivatives of the specific heats. (Changes in the slope of the Mn$_3$Ga curve are marked by arrows.) (b) shows the temperature dependence of the magnetic irreversibility. For better comparison, the differences are normalized to the value at 2 K. [The maxima of the $M(T)$ dependences in Figure 10.12 are marked by vertical lines.]

The magnetic anomalies, which were mentioned in Sec. 10, are not directly visible from the specific heat curves of Mn$_3$Ga or Mn$_2$Ga. This is expected, because the magnetic anomaly covers a wide temperature range rather than appearing as a sharp transition. The situation changes when having a closer look to the derivative of C_v. Figure 10.15(a) compares the differentiated specific heat of Mn$_3$Ga and Mn$_2$Ga to the magnetic irreversibility at low temperatures ($T < 250$ K). The magnetic irreversibility is determined by the difference $m_{FC} - m_{ZFC}$ between the magnetization measured under FC and ZFC conditions. It is evident from Figure 10.15(b) that the magnetic anomaly is more pronounced in the case of Mn$_3$Ga. For Mn$_2$Ga, the curve is flatter than what is expected from the $M(T)$ dependence shown in Figure 10.12. This fact can be addressed to a lowered degree of frustration as noted in Sec. 10. From the comparison of Figures 10.15(a) and (b), it is obvious that the magnetic anomaly coincides with a pronounced change of the slopes of the differentiated specific heat curves. For Mn_3Ga, these changes of the slopes coincide with the strong changes in the magnetic irreversibility. For Mn_2Ga, such changes are much smaller. This is expected from the less pronounced effect that is seen in the temperature dependence of the magnetization.

Summary

The series of compounds $Mn_{3-x}Ga$ was investigated experimentally and by first-principles calculations. In summary, it has been shown that the binary compounds $Mn_{3-x}Ga$, with x varying from 0 to 1.0 crystallize in the tetragonal DO_{22} phase with two inequivalent positions of the Mn atoms. The structural properties were determined by x-ray diffraction. Both the volume and the c/a ratio increase with increasing Mn content. From the electronic structure calculations, this can be attributed to a weakened hybridization between Ga and Mn states along the c axes in the compounds with lower Mn content. With increasing x, mainly - but not exclusively - the Mn_I atoms along the c axis are removed from the tetragonal unit cell. This model is supported by first-principles calculations that showed that the total magnetic moment should increase much stronger with x if only the Mn_I atoms are removed, while it should decrease if both types of Mn atoms are removed simultaneously. The result is an increase of the magnetic moment with x, because the Mn atoms at different crystallographic sites of the $Mn_{3-x}Ga$ compounds order ferrimagnetically. The experimentally observed change of the magnetic moment is also found in the *ab initio* calculations if more Mn_I atoms than Mn_{II} atoms are removed. The coercivity and the different energy products of $Mn_{3-x}Ga$ decrease upon removal of Mn, allowing the magnetic properties and *hardness* to be tuned by varying the composition. The magnetization of the entire series vanishes at approximately 770 K. This is mainly caused by the structural transformation into the hexagonal DO_{19} phase. The entire series exhibits temperature dependent magnetic irreversibilities between zero field-cooled and field-cooled conditions. A magnetic anomaly is found in all samples at temperatures between 145 K and 172 K. This anomaly provides evidence that $Mn_{3-x}Ga$ is a series of magnetically frustrated ferrimagnets. The highest degree of frustration is found for the case of Mn_3Ga, where the magnetic anomaly is most pronounced.

Acknowledgements

The authors thank the staff of the LNLS (Campinas) for support, as well as Fabio Furlan Ferreira and Gustavo Azevedo (LNLS, Campinas) for help with the XRD and

EXAFS measurements. The authors are very grateful to Hubert Ebert and Jan Minar (Munich) for developing and providing the SPRKKR computer code. Financial support by the DFG (Research Unit FOR559), the DAAD (D06/33952), and the CAPES PROBRAL (167/04) is gratefully acknowledged. The authors thank Martin Jourdan for providing the resistance measurements and Gerhard Jakob for helpful discussions (both at the Johannes Gutenberg- Universität, Mainz). Further support of this work was provided by the CNPq, the CT-Energ and by the Brazilian Synchrotron Light Laboratory (LNLS) under proposals D04B - XAFS1 - 5711 and D10B - XPD - 5708.

11 Summary

With the alkali sesquioxides, the Heusler superconductors, and the tetragonal Mn_3Ga, three thematic constellations have been presented within the topic of this work, "Unconventional correlated systems".

The task of the open shell alkali sesquioxides Rb_4O_6 and Cs_4O_6 was to explore their properties experimentally. Extremely challenging syntheses and air sensitivity had made it impossible up to now to characterize these materials. The main focus was on identifying the magnetism of Rb_4O_6 and testing the accuracy of the theoretical prediction of half-metallic ferromagnetism [7]. It was shown in this work that Rb_4O_6 is a magnetically frustrated insulator [8–10]. The magnetization exhibits an irreversibility between ZFC and FC temperature-dependent measurements and an exponential time dependency as known from spin glasses. The related Cs_4O_6 was found to show very similar behavior. The highly molecular character of Rb_4O_6 requires the consideration of electronic correlations in the calculations, and the coexistence of hyperoxide and peroxide anions leads to a reduction of symmetry in the primary cubic crystal. The frustration is due to the peculiar crystal structure of Rb_4O_6, which does not allow a pure antiferromagnetic setup of the hyperoxide anions.

The second job definition in this work was to search for Heusler compounds that exhibit superconductivity according to the van Hove scenario. Certain Heusler compounds with 27 electrons were identified to exhibit saddle points at the L point close to the Fermi energy ϵ_F in their energy dispersion curves. The corresponding high density of states at ϵ_F is a good precondition for superconductivity according to the BCS theory and the van Hove scenario. Five Pd- and Ni-based novel Heusler compounds were synthesized and confirmed to be type II bulk superconductors using resistivity and magnetization measurements. A fact of particular interest is that even the compound $ZrNi_2Ga$ with a large proportion of ferromagnetic Nickel becomes superconducting instead of exhibiting magnetic ordering. This supports the fact that the van Hove scenario, which was already employed to explain superconductivity in the high-T_c superconductors [112], is an important precondition for a compound to show a transition to the highly correlated superconducting state. Noncentrosymmetric superconductors have recently come into the center of attraction, mainly with respect to the task of exploring superconducting gap symmetries and pairing mechanisms. Using the van Hove scenario as a working recipe, the Heusler-type quaternary compound PdAuScAl was identified as an appropriate candidate for noncentrosymmetric superconductivity. The Heusler structure consists of four inter penetrating fcc lattices, and a split of the atoms on the Wyckoff $8c$ position, in this case to $4c$ and $4d$ with Pd and Au on the sites, removes the inversion symmetry and leads to the noncentrosymmetric space group $F\bar{4}3m$. The compound was synthesized and verified to exhibit a superconducting transition at a T_c of 3.0 K. The structural characterization using powder x-ray diffraction revealed, however, that there is a high degree of antisite disorder between Pd and Au, and it is a very interesting question whether this disorder, which changes the symmetry of the structure to $Fm\bar{3}m$ and creates an artificial center of inversion, may be a condition for superconductivity to occur. Attempts of annealing to remove or at least to lower the disorder between Pd and Au failed. Another interesting feature of this compound was found in the specific heat measurement, which showed a second phase transition within the temperature region of the superconducting state. Further experimental investigations are required to explore pairing mechanism and gap symmetry of this compound to evaluate the origin of this second phase transition.

The third setting of a task for this work was the synthesis and characterization of the sample series $Mn_{3-x}Ga$ in the tetragonal ferrimagnetic DO_{22} structure in steps of $x = 0.1$. These metastable phases were found to exhibit hard-magnetic properties up to very high temperatures (700-800 K) [18]. The correlations between itinerant and localized moments of the Mn atoms on the different lattice sites determine a resulting magnetic moment of approximately 1 μ_B per formula unit Mn_3Ga. These magnetic properties make the $Mn_{3-x}Ga$ alloys highly interesting for spin torque transfer applications. Because the magnetic moment increases with x whereas the magnetism becomes softer with increasing x, a continuous tunability of the magnetic properties of these alloys is easily possible by adjusting the Mn-content to the desired materials properties.

One major aspect if not the most important one in the research area of materials science is the preselection of prospective candidates for a desired technological application. Only accurate assessments of the microscopic properties of a material such as its electronic structure can lead to the desired macroscopic properties and thus to successful and rational material design. In this context it is of particular importance to understand and evaluate electronic correlations to predict and verify macroscopic properties of materials. In 1963 Slater defined correlation as the energy difference between the Hartree-Fock energy and the experimentally observed energy [158]. He furthermore states that "it is intuitively obvious that two electrons, repelling each other by Coulomb repulsion, will be less likely to be found at the same point of space than at points separated from each other." Ever-growing computer capacities will hopefully contribute to a complete understanding of electronic correlations in the near future so that the energy between the theoretical model and the experiment will finally converge to zero.

Bibliography

[1] J. G. Bednorz and K. A. Müller. *Z. Phys.*, B64:189, 1986.

[2] Y. Kamihara, T. Watanabe, M. Hirano, and H. Hosono. *J. Am. Chem. Soc.*, 130:3296, 2008.

[3] G. H. Jonker and J. H. Van Santen. *Physica*, 16:377, 1950.

[4] R. von Helmolt, J. Wecker, W. Holzapfel, L. Schultz, and K. Samwer. *Phys. Rev. Lett.*, 71:2331, 1993.

[5] R. von Helmolt, J. Wecker, W. Holzapfel, L. Schultz, and K. Samwer. *Phys. Rev. Lett.*, 53:2339, 1984.

[6] M. Sigrist and K. Ueda. *Rev. Mod. Phys.*, 63:239, 1991.

[7] J. J. Attema, G. A. de Wijs, G. R. Blake, and R. A. de Groot. *J. Am. Chem. Soc.*, 127:16325, 2005.

[8] J. Winterlik, G. H. Fecher, C. Felser, C. Mühle, and M. Jansen. *J. Am. Chem. Soc.*, 129:6990, 2007.

[9] J. Winterlik, G. H. Fecher, C. A. Jenkins, C. Felser, C. Mühle, K. Doll, M. Jansen, L. M. Sandratskii, and J. Kübler. *Phys. Rev. Lett.*, 102:016401, 2009.

[10] J. Winterlik, G. H. Fecher, C. A. Jenkins, C. Felser, C. Mühle, M. Jansen, T. Palasyuk, I. Trojan, S. Medvedev, M. I. Eremets, and F. Emmerling. *Submitted to Phys. Rev. B*, 2009.

[11] J. Winterlik, G. H. Fecher, and C. Felser. *Solid State Commun.*, 145:475, 2008.

[12] J. Winterlik, G. H. Fecher, C. Felser, M. Jourdan, K. Grube, F. Hardy, H. von Löhneysen, K. L. Holman, and R. J. Cava. *Phys. Rev. B*, 78:184506, 2008.

[13] J. Winterlik, G. H. Fecher, A. Thomas, and C. Felser. *Phys. Rev. B*, 79:064508, 2009.

[14] L. van Hove. *Phys. Rev.*, 89:1189, 1953.

[15] C. Felser. *J. Sol. State Chem.*, 160:93, 2001.

[16] S. Wurmehl, H. C. Kandpal, G. H. Fecher, and C. Felser. *J. Phys.: Condens. Matter*, 18:6171, 2006.

[17] B. Balke, G. H. Fecher, J. Winterlik, and C. Felser. *Appl. Phys. Lett.*, 90:152504, 2007.

[18] J. Winterlik, B. Balke, G. H. Fecher, and C. Felser. *Phys. Rev. B*, 77:054406, 2008.

[19] F. Wu, S. Mizukami, D. Watanabe, H. Naganuma, M. Oogane, Y. Ando, and T. Miyazaki. *Appl. Phys. Lett.*, 94:122503, 2009.

[20] M. Jansen and N. Korber. *Z. Anorg. Allg. Chem.*, 589/599:163, 1991.

[21] A. Helms and W. Klemm. *Z. Anorg. Allg. Chemie*, 242:201, 1939.

[22] M. Jansen, R. Hagenmayer, and N. Korber. *C. R. Acad. Sci.*, Ser. IIc: Chim.:591, 1999.

[23] N. Korber, W. Assenmacher, and M. Jansen. *Prax. Naturwiss. Chem.*, 40:18, 1991.

[24] A. P. Ramirez. *Annu. Rev. Mater. Sci.*, 24:453, 1994.

[25] S. M. Humphrey, A. Alberola, C. J. Gómez García, and P. T. Wood. *Chem. Commun.*, 15:1607, 2006.

[26] J. Winterlik, G. H. Fecher, C. Felser, C. Mühle, and M. Jansen. *Unpublished work*, 2007.

[27] G. Brauer. *Handbuch der Präparativen Anorganischen Chemie*, volume 2. Enke-Verlag: Stuttgart, Germany, 1978.

[28] M. Labhart, D. Raoux, W. Känzig, and M. A. Bösch. *Phys. Rev. B*, 20:53, 1979.

[29] J. J. Attema, G. A. de Wijs, and R. A. de Groot. *J. Phys.: Condens. Matter*, 19:165203, 2007.

[30] R. B. Meier and R. B. Helmholdt. *Phys. Rev. B*, 29:1387, 1984.

[31] I. N. Goncharenko, O. L. Makarova, and L. Ulivi. *Phys. Rev. Lett.*, 93:055502–1, 2004.

[32] Y. Akahama, H. Kawamura, D. Husermann, M. Hanfland, and O. Shimomura. *Phys. Rev. Lett.*, 74:4690, 1995.

[33] K. Shimizu, K. Suhara, M. Ikumo, M. I. Eremets, and K. Amaya. *Nature*, 393:767, 1998.

[34] I. A. Nekrasov, M. A. Korotin, and V. I. Anisimov. *arXiv:cond-mat/0009107v1.*, 2000.

[35] A. Simon. *Z. Anorg. Allg. Chemie*, 431:5, 1977.

[36] W. Klein, K. Armbruster, and M. Jansen. *Chem. Comm.*, 6:707, 1998.

[37] Y.-W. Son, M. L. Cohen, and S. G. Louie. *Nature*, 444:347, 2006.

[38] I. S. Elfimov, A. Rusydi, S. I. Csiszar, Z. Hu, H. H. Hsieh, H.-J. Lin, C. T. Chen, R. Liang, and G. A. Sawatzky. *Phys. Rev. Lett.*, 98:137202, 2007.

[39] K. Binder and A. P. Young. *Rev. Mod. Phys.*, 58:801, 1986.

[40] C. Urano, M. Nohara, S. Kondo, F. Sakai, H. Takagi, T. Shiraki, and T. Okubo. *Phys. Rev. Lett.*, 85:1052, 2000.

[41] G. Khaliullin, P. Horsch, and A. M. Oles. *Phys. Rev. Lett.*, 86:3879, 2001.

[42] T. Bremm and M. Jansen. *Z. anorg. allg. Chemie*, 610:64, 1992.

[43] H. Seyeda and M. Jansen. *J. Chem. Soc.*, Dalton Trans.:165203, 1998.

[44] R. Dovesi, V. R. Saunders, C. Roetti, R. Orlandeo, C. M. Zicovich-Wilson, F. Pascale, B. Civalleri, K. Doll, N. M. Harrison, I. J. Bush, Ph. D'Arco, and M. LLunell. *CRYSTAL 2006 User's Manual*. University of Torino, Torino, 2006.

[45] http://www.aip.org/pubservs/epaps.html. See EPAPS Document No. E-PRLTAO-102-007902 for details about the methodologies of the electronic structure calculations used in this work. For more information on EPAPS.

[46] A. D. Becke. *J. Chem. Phys.*, 98:5648, 1993.

[47] A. R. Williams, J. Kübler, and C. D. Gelatt. *Phys. Rev. B*, 19:6094, 1979.

[48] V. I. Anisimov, J. Zaanen, and O. K. Andersen. *Phys. Rev. B*, 44:943, 1991.

[49] M. Uhl, L. M. Sandratskii, and J. Kübler. *Phys. Rev. B*, 50:291, 1994.

[50] C. Cao, S. Hill, and H.-P. Cheng. *Phys. Rev. Lett.*, 100:167206, 2008.

[51] L. M. Sandratskii. *Adv. Phys.*, 47:91, 1998.

[52] Y. A. Freiman and H. J. Jodl. *Phys. Reports*, 401:1, 2004.

[53] W. Hesse, M. Jansen, and W. Schnick. *Prog. Solid State Chem.*, 19:47, 1989.

[54] A. Helms and W. Klemm. *Z. Anorg. Allg. Chemie*, 242:33, 1939.

[55] A. Helms and W. Klemm. *Z. Anorg. Allg. Chemie*, 242:97, 1939.

[56] S. Kuroiwa, Y. Saura, J. Akimitsu, M. Hiraishi, M. Miyazaki, K. H. Satoh, S. Takeshita, and R. Kadono. *Phys. Rev. Lett.*, 100:097002, 2008.

[57] L. Hackspill. *Helv. Chim. Acta*, 11:1008, 1928.

[58] H. H. Eysel and S. Thym. *Z. anorg. allg. Chem.*, 411:97, 1975.

[59] J. B. Bates, M. H. Brooker, and G. E. Boyd. *Chem. Phys. Lett.*, 16:391, 1972.

[60] Jörg Schmalian and Peter G. Wolynes. *Phys. Rev. Lett.*, 85:836, 2000.

[61] R. S. Keizer, S. T. B. Goennewein, T. M. Klapwijk, G. Miao, G. Xiao, and A. Gupta. *Nature*, 439:825, 2006.

[62] H. C. Kandpal, G. H. Fecher, and C. Felser. *J. Phys. D: Appl. Phys.*, 40:1507, 2007.

[63] C. Felser, G. H. Fecher, and B. Balke. *Angew. Chem. Int. Ed.*, 46:668, 2007.

[64] T. Marukame, T. Ishikawa, S. Hakamata, K. Matsuda, T. Uemura, and M. Yamamoto. *Appl. Phys. Lett.*, 90:012508, 2007.

[65] N. Tezuka, N. Ikeda, S. Sugimoto, and K. Inomata. *Appl. Phys. Lett.*, 89:252508, 2006.

[66] M. Ishikawa, J. L. Jorda, and A. Junod. *Superconductivity in d- and f-band metals 1982*. W. Buckel and W. Weber, Kernforschungszentrum Karlsruhe, Germany, 1982.

[67] J. H. Wernick, G. W. Hull, J. E. Bernardini, and J. V. Waszczak. *Mater. Lett.*, 2:90, 1983.

[68] H. A. Kierstead, B. D. Dunlap, S. K. Malik, A. M. Umarji, and G. K. Shenoy. *Phys. Rev. B*, 32:135, 1985.

[69] R. N. Shelton, L. S. Hausermann-Berg, M. J. Johnson, P. Klavins, and H. D. Yang. *Phys. Rev. B*, 34:199, 1986.

[70] G. Gladstone, M. A. Jensen, and J. R. Schrieffer. *Superconductivity Vol. 2*. Parks, R. D., USA, 1969.

[71] B. T. Matthias. *Phys. Rev.*, 92:874, 1953.

[72] S. V. Vonsovsky, Y. A. Izyumov, and E. Z. Kurmaev. *Superconductivity of Transition Metals*. Springer-Verlag, Berlin, Heidelberg, 1982.

[73] E. R. Schrader, N. S. Freedman, and J. C. Faken. *Appl. Phys. Lett.*, 4:105, 1964.

[74] P. Blaha, K. Schwarz, G. K. H. Madsen, D. Kvasnicka, and J. Luitz. *WIEN2k, An Augmented Plane Wave + Local Orbitals Program for Calculating Crystal Properties*. Karlheinz Schwarz, Techn. Universitaet Wien, Wien, Austria, 2001.

[75] J. P. Perdew, K. Burke, and M. Ernzerhof. *Phys. Rev. Lett.*, 77:3865, 1996.

[76] D. M. Teter, G. V. Gibbs, Jr. M. B. Boisen, D. C. Allan, and M. P. Teter. *Phys. Rev. B*, 52:8064, 1995.

[77] S. Wurmehl, G. H. Fecher, H. C. Kandpal, V. Ksenofontov, C. Felser, and H.-J. Lin. *Appl. Phys. Lett.*, 88:032503, 2005.

[78] F. Heusler. *Verhandl. DPG*, 5:219, 1903.

[79] P. Blaha, K. Schwarz, P. Sorantin, and S.B. Tricky. *Comput. Phys. Commun.*, 59:399, 1990.

[80] K. Parlinski. *Software Phonon*, 2006.

[81] H. Ebert. Electronic structure and physical properties of solids - fully relativistic band structure calculations for magnetic solids - formalism and application. In H. Dreysse, editor, *Lecture Notes in Physics*, volume 535, pages 191–246. Springer Verlag, 1999.

[82] H. Ebert. *The Munich SPR-KKR package, Version 3.6, http://olymp.cup.uni-muenchen.de/ak/ebert/SPRKKR*. 2005.

[83] S. H. Vosko and L. Wilk. *Phys. Rev. B*, 22:3812, 1980.

[84] S. H. Vosko, L. Wilk, and M. Nusair. *Can. J. Phys.*, 58:1200, 1980.

[85] P. Mohn, P. Blaha, and K. Schwarz. *J. Magn. Magn. Mater.*, 140-144:183 – 184, 1995.

[86] S. C. Lee, T. D. Lee, P. Blaha, and K. Schwarz. *J. Appl. Phys.*, 97:10C307, 2005.

[87] J. P. Perdew, A. Ruzsinszky, G. I. Csonka, O. A. Vydrov, G. E. Scuseria, L. A. Constantin, X. Zhou, and K. Burke. *Phys. Rev. Lett.*, 100:136406, 2008.

[88] C. Bungaro, K. M. Rabe, and A. Dal Corso. *Phys. Rev. B*, 68:134104, 2003.

[89] A. T. Zayak, P. Entel, K. M. Rabe, W. A. Adeagbo, and M. Acet. *Phys. Rev. B*, 72:054113, 2005.

[90] W. H. Butler and G. M. Stocks. *Phys. Rev. B*, 29:4217, 1984.

[91] P. Weinberger. *Electron Scattering Theory for Ordered and Disordered Matter*. Clarendon Press; Oxford, 1990.

[92] H. Gutfreund, M. Weger, and O. Entin-Wohlman. *Phys. Rev. B*, 31:606, 1985.

[93] P. W. Anderson. *J. Phys. Chem. Solids*, 11:26, 1959.

[94] M. Ma and P. A. Lee. *Phys. Rev. B*, 32:5658, 1985.

[95] S. Waki, Y. Yamaguchi, and K. Mitsugi. *J. Phys. Soc. Jpn.*, 54:1673, 1985.

[96] Y. Kamihara, H. Hiramatsu, M. Hirano, R. Kawamura, H. Yanagi, T. Kamiya, and H. Hosono. *J. Am. Chem. Soc.*, 128:10012, 2006.

[97] V. Hlukhyya, N. Chumalo, V. Zaremba, and T. F. Fässler. *Z. Anorg. Allg. Chem.*, 634:1249, 2008.

[98] J. Labbé and J. Friedel. *J. Phys. (Paris)*, 27:153, 1966.

[99] K. Schwarz, P. Blaha, and G. K. H. Madsen. *Comput. Phys. Commun.*, 147:71, 2002.

[100] A. T. Zayak and P. Entel. *J. Magn. Magn. Mat.*, 290-291:874, 2005.

[101] H. Padamsee, J. E Neighbor, and C. A. Shiffman. *J. Low Temp. Phys.*, 12:387, 1973.

[102] B. Mühlschlegel. *Z. Phys.*, 155:313, 1959.

[103] N. R. Werthamer, E. Helfand, and P.C. Hohenberg. *Phys. Rev.*, 147:295, 1966.

[104] A. M. Clogston. *Phys. Rev. Lett.*, 9:266, 1962.

[105] F.S. da Rocha, G.L.F. Fraga, D.E. Brandao, C.M. da Silva, and A.A. Gomes. *Physica B*, 269:154, 1999.

[106] M. A. S. Boff, G. L. F. Fraga, D. E. Brandao, A. A. Gomes, and T. A. Grandi. *Phys. Stat. Sol. (a)*, 154:549, 1996.

[107] W. L. McMillan. *Phys. Rev.*, 167:331, 1968.

[108] W. Lin and A. J. Freeman. *Phys. Rev. B*, 45:61, 1991.

[109] M. A. S. Boff, G. L. F. Fraga, D. E. Brandao, and A. A. Gomes. *J. Mag. Magn. Mat.*, 153:135, 1996.

[110] A. Simon, A. Yoshiasa, M. Bäcker, R. W. Henn, C. Felser, R. K. Krenzer, and Mattausch H. *Z. Anorg. Allg. Chem.*, 622:123, 1996.

[111] A. Simon. *Angew. Chem. Int. Ed. Engl.*, 36:1788, 1997.

[112] A. Simon. *Angew. Chem.*, 99:602, 1987.

[113] A. Abrikosov, J. C. Campuzano, and Gofron K. *Phys. Status Solidi C*, 214:73, 1993.

[114] D. M. News, C. C. Tsuei, and P. C. Pattniak. *Phys. Rev. B*, 52:13611, 1995.

[115] A. Simon. *Angew. Chem.*, 109:1872, 1997.

[116] E. Bauer, G. Hilscher, H. Michor, C. Paul, E. W. Scheidt, A. Gribanov, Y. Seropegin, H. Noël, M. Sigrist, and P. Rogl. *Phys. Rev. Lett.*, 92:027003, 2004.

[117] N. Kimura, K. Ito, K. Saitoh, Y. Umeda, H. Aoki, and T. Terashima. *Phys. Rev. Lett.*, 95:247004, 2005.

[118] I. Sugitani, Y. Okuda, H. Shishido, T. Yamada, A. Thamizhavel, E. Yamamoto, T. D. Matsuda, Y. Haga, T. Takeuchi, R. Settai, and Y. Onuki. *J. Phys. Soc. Jpn.*, 75:043703, 2006.

[119] T. Akazawa, H. Hidaka, T. Fujiwara, T. C. Kobayashi, E. Yamamoto, Y. Haga, R. Settai, and Y. Onuki. *J. Phys.: Condens. Matter*, 16:L29, 2004.

[120] G. Amano, S. Akutagawa, T. Muranaka, Y. Zenitani, and J. Akimitsu. *J. Phys. Soc. Jpn.*, 73:530, 2004.

[121] K. Togano, P. Badica, Y. Nakamori, S. Orimo, H. Takeya, and K. Hirata. *Phys. Rev. Lett.*, 93:247004, 2004.

[122] P. Badica, T. Kondo, and K. Togano. *J. Phys. Soc. Jpn.*, 74:1014, 2005.

[123] T. Klimczuk, F. Ronning, V. Sidorov, R. J. Cava, and J. D. Thompson. *Phys. Rev. Lett.*, 99:257004, 2007.

[124] G. Schuck, S. M. Kazakov, K. Rogacki, N. D. Zhigadlo, and J. Karpinski. *Phys. Rev. B*, 73:144506, 2006.

[125] P. A. Frigeri, D. F. Agterberg, A. Koga, and M. Sigrist. *Phys. Rev. Lett.*, 92:097001, 2004.

[126] H. Q. Yuan, D. F. Agterberg, N. Hayashi, P. Badica, D. Vandervelde, K. Togano, M. Sigrist, and M. B. Salamon. *Phys. Rev. Lett.*, 97:017006, 2006.

[127] E. Engel and S. H. Vosko. *Phys. Rev. B*, 47:13164, 1993.

[128] C. Pfleiderer, J. Bœuf, and H. v. Löhneysen. *Phys. Rev. B*, 65:172404, 2002.

[129] C. Pfleiderer. *Physica B*, 329:1085, 2003.

[130] S. Tomiyoshi, E. R. Cowley, , and H. Onodera. *Phys. Rev. B*, 73:024416, 2006.

[131] G. Kadar and E. Kren. *Int. J. Magn.*, 1:143, 1971.

[132] S. Tomiyoshi, Y. Yamaguchi, and T. Nagamiya. *J. Magn. Magn. Mater.*, 31:629, 1983.

[133] J. W. Cable, N. Wakabayashi, and P. Radhakrishna. *Phys. Rev. B*, 48:6159, 1993.

[134] H. Takizawa, T. Yamashita, K. Uheda, and T. Endo. *J. Phys.: Cond. Mat.*, 14:11147, 2002.

[135] H. Ohmori, S. Tomiyoshi, H. Yamauchi, and H. Yamamoto. *J. Magn. Magn. Mater.*, 70:249, 1987.

[136] J. W. Cable, N. Wakabayashi, and P. Radhakrishna. *Solid State Comm.*, 88:161, 1993.

[137] T. Yamashita, H. Takizawa, T. Sasaki, K. Uheda, and T. Endo. *J. Alloys Compds.*, 348:220, 2003.

[138] V. S. Goncharov and V. M. Ryzhkovskii. *Inorg. Mater.*, 41:557, 2005.

[139] M. Ellner. *J. Appl. Crys.*, 13:99, 1980.

[140] A. Bekhti-Siad, A. Mokrani, C. Demangeat, and A. Khelil. *J. Mol. Struct.: THEOCHEM*, 777:11, 2006.

[141] M. Suzuki, M. Shirai, and K. Motizuki. *J. Phys.: Cond. Mat.*, 4:L33, 1992.

[142] V. M. Ryzhkovskii, V. P. Glazkov, V. S. Goncharov, D. P. Kozlenko, and B. N. Savenko. *Phys. Solid State*, 44:2281, 2002.

[143] I. Tsuboya and M. Sugihara. *J. Phys. Soc. Japan*, 18:143, 1963.

[144] M. Hasegawa and I. Tsuboya. *J. Phys. Soc. Japan*, 20:464, 1965.

[145] H.-G. Meißner and K. Schubert. *Zeitschrift für Metallkunde*, 56:523, 1965.

[146] S. Wurmehl, H. C. Kandpal, G. H. Fecher, and C. Felser. *J. Phys.: Condens. Matter*, 18:6171 – 6181, 2006.

[147] E. Kren and G. Kadar. *Solid State Comm.*, 8:1653 – 1655, 1970.

[148] H. Niida, H. Hori, T. Onodera, Y. Yamaguchi, and Y. Nakagawa. *J. Appl. Phys.*, 79:5946, 1996.

[149] J. Kübler. *J. Phys.: Condens. Matter*, 18:9795 – 9807, 2006.

[150] H. Ebert. Electronic structure and physical properties of solids - fully relativistic band structure calculations for magnetic solids - formalism and application. In H. Dreysse, editor, *Lecture Notes in Physics*, volume 535, pages 191–246. Springer Verlag, 1999.

[151] H. C. N. Tolentino, A. Y. Ramos, M. C. M. Alves, R. A. Barrea, E. Tamura, J. C. Cezar, and N. Watanabe. *J. Synchrotron Radiat.*, 8:1040, 2001.

[152] M. Newville. *J. Synchrotron Radiat.*, 8:322, 2001.

[153] S. I. Zabinsky, J. J. Rehr, A. Ankudinov, R. C. Albers, and M. J. Eller. *Phys. Rev. B*, 52:2995, 1995.

[154] I. Mirebeau, G. Iancu, M. Hennion, G. Gavoille, and J. Hubsch. *Phys. Rev. B*, 54:15928, 1996.

[155] P. J. Ford, E. Babic, and J. A. Mydosh. *J. Phys. F: Metal Phys.*, 3:L75, 1973.

[156] I. A. Campbell, P. J. Ford, and A. Hamzic. *Phys. Rev. B*, 26:5195, 1982.

[157] M. Escorne, A. Mauger, J. L. Tholence, and R. Triboulet. *Phys. Rev. B*, 29:6306, 1984.

[158] J. C. Slater. *Rev. Mod. Phys.*, 35:484, 1963.

I want morebooks!

Buy your books fast and straightforward online - at one of the world's fastest growing online book stores! Environmentally sound due to Print-on-Demand technologies.

Buy your books online at
www.get-morebooks.com

Kaufen Sie Ihre Bücher schnell und unkompliziert online – auf einer der am schnellsten wachsenden Buchhandelsplattformen weltweit! Dank Print-On-Demand umwelt- und ressourcenschonend produziert.

Bücher schneller online kaufen
www.morebooks.de

OmniScriptum Marketing DEU GmbH
Heinrich-Böcking-Str. 6-8
D - 66121 Saarbrücken
Telefax: +49 681 93 81 567-9

info@omniscriptum.com
www.omniscriptum.com

Printed by Books on Demand GmbH, Norderstedt / Germany